Desert Farmers at the River's Edge

The Hohokam and Pueblo Grande

by John P. Andrews and Todd W. Bostwick

Pueblo Grande Museum and Archaeological Park

About the Authors

John P. Andrews is an archaeologist serving as the Coordinator of Public Programs at Pueblo Grande Museum and Archaeological Park. He is responsible for all educational and public programs at the Museum. John received his B.S. in Anthropology and a certificate in Museum Studies from Central Michigan University. His interests in archaeology include Great Lakes archaeology, Southwest archaeology, and public education.

Todd W. Bostwick is the Phoenix City Archaeologist. He is responsible for the management of archaeological sites within the city and is the editor of the Pueblo Grande Museum Anthropological Papers. Todd has a B.A. in Anthropology from the University of Nevada-Reno and an M.A. in Anthropology from Arizona State University. His interests include Hohokam rock art, archaeoastronomy, ideology, craft production, and ruins stabilization.

Desert Farmers at the River's Edge

The Hohokam and Pueblo Grande

John P. Andrews
and
Todd W. Bostwick

Pueblo Grande Museum and Archaeological Park
A National Historic Landmark

Second Edition.
Revised and enlarged.
2000

City of Phoenix
Parks, Recreation and
Library Department

Phoenix, Arizona

Pueblo Grande Museum and Archaeological Park
4619 East Washington Street
Phoenix, Arizona 85034-1909
(602) 495-0901
www.pueblogrande.com

Pueblo Grande Museum is
Accredited by the American Association of Museums

City of Phoenix
Parks, Recreation and Library Department

The mission of the Parks, Recreation and Library Department is to enhance the quality of life by providing and maintaining the richness and diversity of an environmentally sensitive, safe, available, and affordable parks, recreation and library system that fulfills the physical, intellectual, mental, social, cultural and leisure needs of all people.

Library of Congress Catalog Card Number: 99-74031
ISBN 1-882572-30-0

Printed in the United States.

Table of Contents

Preface

Desert Farmers at the River's Edge: The Hohokam and Pueblo Grande is about the people who once lived at the prehistoric village known today as Pueblo Grande. The story of these people, called the Hohokam, is presented here for a general audience. Characteristics of the Hohokam culture are briefly discussed, and specific information about Pueblo Grande is summarized.

Topics discussed in this book include Hohokam cultural origins, environment, subsistence, domestic architecture, monumental and public architecture, material culture, trade networks, ideology, and cultural collapse. Integrated throughout each topic is information regarding archaeological evidence from Pueblo Grande.

The adobe and stone ruins located at 4619 East Washington Street in Phoenix, Arizona have been recognized as a prehistoric village since the 1860s. Archaeologists have probed these buried deposits since the 1880s, and the City of Phoenix has preserved the central portion of the site since 1924. In the mid-1960s, two portions of the modern-day park, the Pueblo Grande platform mound and Park of Four Waters were nominated to the National Register of Historic Places.

Much of what is now known about Pueblo Grande is new information resulting from three recent archaeological investigations. One of these projects is an ongoing archival study of unpublished excavation data, collected from 1929 to 1981, from the City of Phoenix portion of the Pueblo Grande site. This 102 acre area comprises the Pueblo Grande Museum and Archaeological Park. The second project consisted of the Arizona Department of Transportation (ADOT) sponsored excavation in 1989 and 1990 within the right-of-way of State Route 143, known as the Hohokam Expressway. This project, which included the eastern portion of the prehistoric Pueblo Grande village site, but outside today's archaeological park boundaries, was undertaken by Soil Systems, Inc., an archaeological consulting firm. In 1999 and 2000, Soil Systems, Inc. again conducted major excavations, this time to the north of the park boundaries.

More detailed information about the Hohokam and Pueblo Grande can be obtained from the many research reports generated from these projects. See the *Recommended Reading* list at the end of this book for a selection of general and technical books about the Hohokam and the archaeological site of Pueblo Grande. This book was produced by Pueblo Grande Museum and printed by the City of Phoenix Printing Services, City Clerk Department.

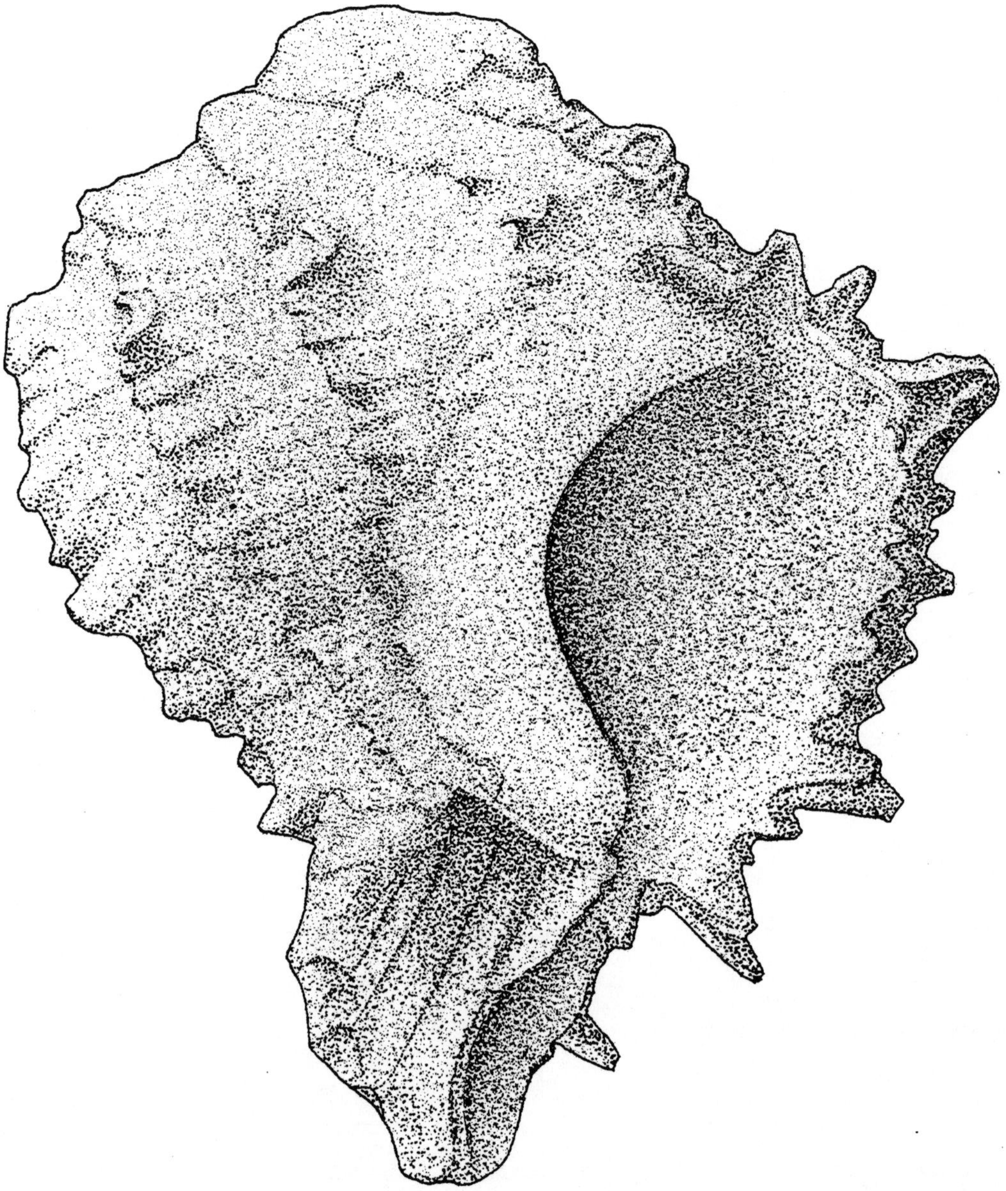

Miracanthus shell trumpet found at Pueblo Grande (actual size). Shell trumpets similar to this one have been found at other Hohokam sites such as Snaketown and Casa Grande. The end of this trumpet was ground off to form a type of mouthpiece for blowing. Shell trumpets may have been used to call villagers to the platform mound.

The First Americans of the Southwest

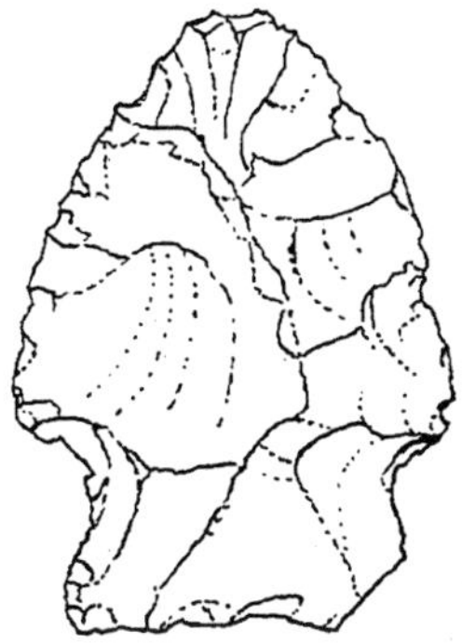

Archaic style dart point from the Pueblo Grande platform mound.

The exact date that American Indians first occupied the Americas is open to debate. Archaeological evidence suggests that sometime around 14,000 years ago or earlier, the ancestors of American Indians arrived on the North American continent. Archaeologists generally believe that American Indians migrated from northeast Asia across the Bering Straits land bridge, known as *Beringia*, between Siberia and Alaska during the late *Pleistocene Epoch*, or last Ice Age. Over the course of centuries, these new arrivals settled both the North and South American continents from Alaska to Tierra del Fuego on the tip of South America.

The first American Indians, called *Paleo-Indians* (paleo meaning "ancient") by archaeologists, were hunters and gatherers. They hunted Pleistocene *megafauna*, or big game animals, such as mammoths, mastodons, giant bison, ancient horses, camels, and giant sloths. Although mammoths and other Pleistocene mammal skeletons have been discovered in the Salt River Valley, no Paleo-Indian remains have yet been recovered. However, several mammoth kill sites containing Paleo-Indian spear points have been excavated in southeastern Arizona.

With the end of the Pleistocene, a warmer climate resulted in mass extinctions of Ice Age megafauna. Noted paleobotanist Paul Martin has proposed that Paleo-Indians may have hunted the megafauna into extinction, but current data indicate major climatic changes had the greatest influence. With the disappearance of most of the megafauna, the Paleo-Indians were forced to hunt smaller game, which led to the development of a new American Indian culture archaeologists call the *Archaic* (meaning "old").

Sometime around 9,000 years ago, Desert Archaic people lived throughout the American Southwest. These small bands of hunters and gatherers led a nomadic life, traveling from place to place with the seasons, searching out a variety of wild plants. About 3,000 years ago their way of life began to dramatically change as many Archaic peoples of the Southwest adopted an agricultural lifestyle and became more sedentary. As time went on and farming became more established, groups began developing differences in their material culture. Through these differences, cultures of the Southwest became more visibly distinct from one another.

Archaeologists have defined several prehistoric cultures in the Southwest. In the Four Corners area of Utah, Colorado,

New Mexico and Arizona, groups known today as the *Anasazi* (ancestral Pueblo) began constructing permanent hamlets and villages. The *Mogollon* people lived along the rim of the Colorado Plateau and the mountains of eastern Arizona and western New Mexico. The *Patayan* occupied the area along the Colorado River in western Arizona. Finally, the *Hohokam,* (pronounced ho-ho-KHAM), settled large villages along the rivers of central and southern Arizona.

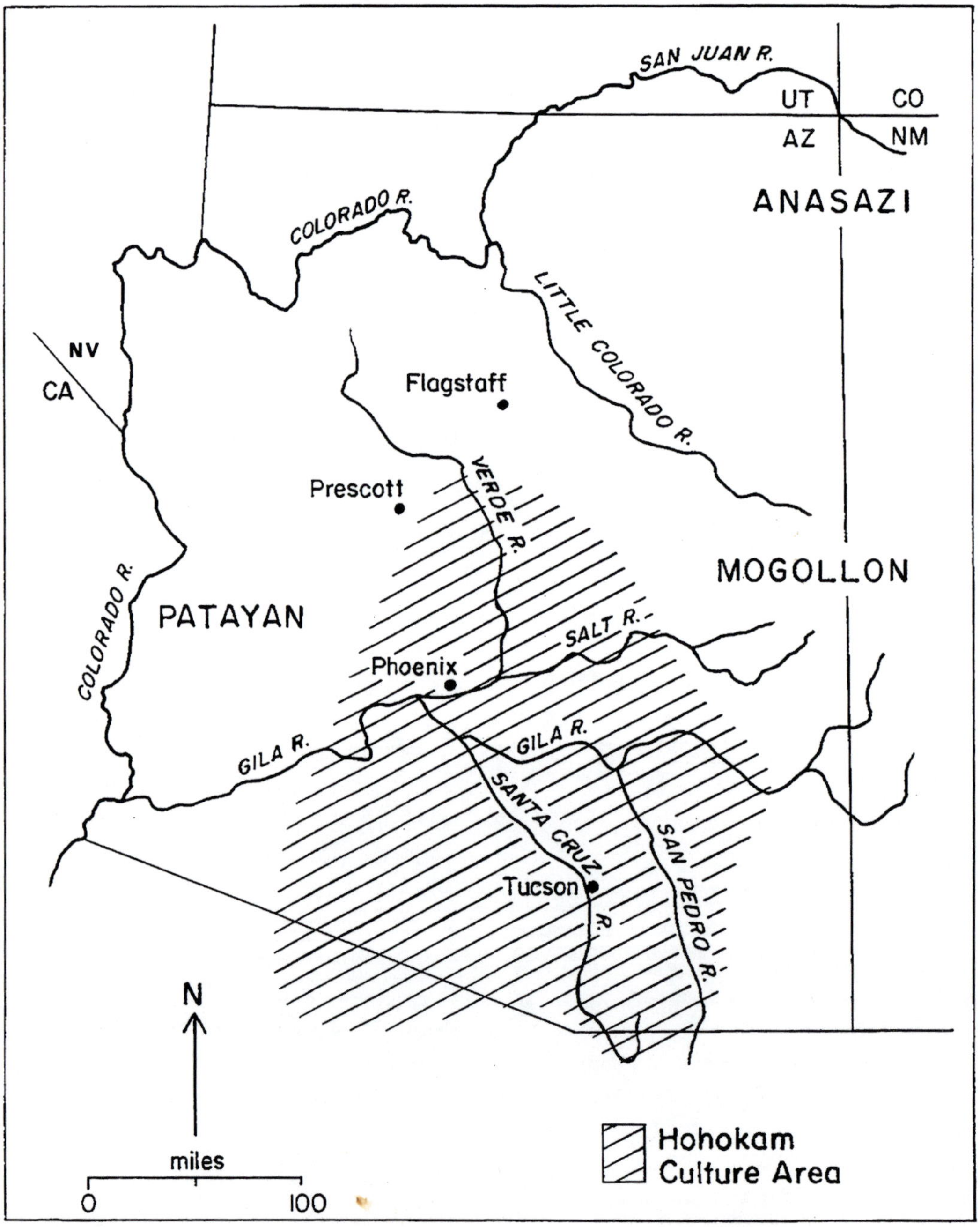

The Hohokam culture area and neighboring cultures of the American Southwest.

The Advent of the Hohokam

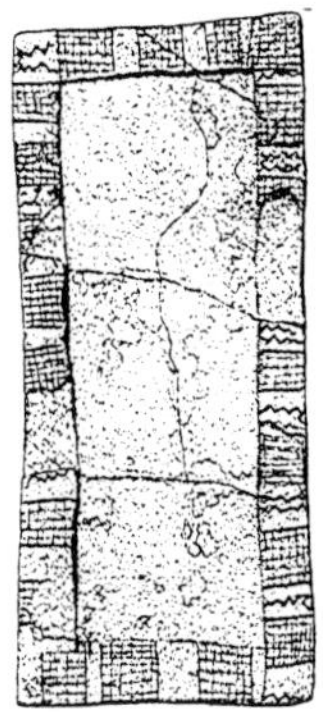

Tabular schist palette from Pueblo Grande.

The Hohokam were a farming people who lived in south-central Arizona and northern Mexico from approximately A.D. 1 to 1450. Their name comes from the Pima Indian (Akimel O'odham) word for "those who have gone" or "all used up." A creative and industrious people, the Hohokam turned the arid desert of the Salt and Gila river valleys and other areas of southern Arizona into fertile farmlands and thriving villages. One of those villages is the site of Pueblo Grande, preserved as a City of Phoenix museum and archaeological park.

Ruins containing the Hohokam's characteristic red-on-buff pottery are scattered over an area of at least 27,000 square miles (69,930 sq. km), but are most frequent in the Salt River Valley and the nearby Gila River Valley. These desert farmers are admired for their lengthy occupation of the Sonoran Desert and their impressive cultural accomplishments in architecture, crafts, and irrigation agriculture. In addition, the Hohokam were active in trade networks with other prehistoric cultures that stretched across the entire Southwest, into Mexico, and to the Pacific Coast. However, where did the Hohokam come from?

Cultural Origins

The origin of the Hohokam culture is still debated among archaeologists. It was once theorized that the Hohokam were Mesoamerican immigrants or traders who came north into south-central Arizona out of Mexico. Although there is little doubt that Mexican cultures influenced the Hohokam, many archaeologists today argue that the Hohokam culture arose out of local hunting and gathering groups of the Desert Archaic culture who had lived in the Southwest since around 7000 B.C. Because the Hohokam were well adapted to the challenges of the Sonoran Desert environment, and many material cultural similarities with the Desert Archaic remained part of the Hohokam culture for generations, it seems unlikely that they immigrated from elsewhere. In essence, the Hohokam culture can be seen as a new culture emerging out of their Archaic predecessors.

Around 1500 B.C., knowledge of agriculture spread into the Southwest out of Mexico where farming was already well developed for thousands of years. Crops included corn (maize), beans, squash and cotton. With this knowledge of farming, the ancestors of the Hohokam, hunters and gatherers who had once moved across the landscape with the seasons, settled down into permanent villages and began practicing agriculture in the desert.

Besides the knowledge of agriculture, other ideas also traveled north out of Mexico and influenced the Hohokam. From Mexican cultures, the Hohokam may have learned ceramic manufacturing techniques and borrowed elements of their religious ideology.

Yet who were the Hohokam people? Did they all speak the same language? Were they a single group of people who spread out to colonize the desert, or were they several different ethnic groups who shared many cultural traits? From the archaeological record, the answers to these questions are uncertain. Archaeologists have little way of knowing if the Hohokam were a single ethnic group with a single language. It may be that the Hohokam did not all have the same ethnic background but were a people who shared a common material culture, language(s), ideology and general lifestyle.

Hohokam Chronology

Archaeologists have documented a number of changes in Hohokam culture through time. These cultural changes, primarily relating to pottery styles and architecture, have resulted in the Hohokam cultural sequence being separated into four major periods, subsequently divided into phases. The dates and characteristics of those periods and phases are still debated, and remain one of the major challenges of Hohokam archaeology. For an overview of the general characteristics of Hohokam culture and major events at the village of Pueblo Grande, see the table (page 5) on the *Hohokam Cultural Sequence*.

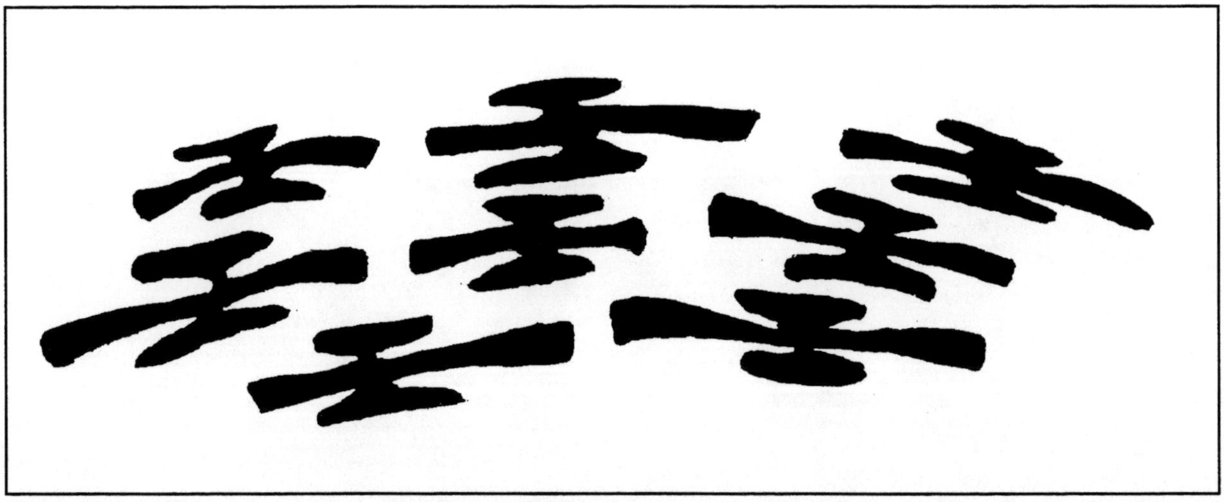

Flying bird motif found on Hohokam pottery at Pueblo Grande and other Hohokam sites. During the Colonial period (ca. A.D. 750-900), the Hohokam painted various life forms on their pottery. During later periods, geometric designs came more into fashion and life forms were not used.

HOHOKAM CULTURAL SEQUENCE

Year	Period	Phase	General Hohokam Characteristics	Pueblo Grande Village
1540	**Historic**		Spanish explore southern Arizona. Development of historic Pima (Akimel O'odham) and Tohono O'odham (Papago).	
1450	**Protohistoric** (1450-1540)		Hohokam cultural collapse.	Pueblo Grande abandoned.
1350		*Polvoron*	Pit structures built. Population decline.	Architectural styles change.
1300		*Civano*	Salado Polychrome pottery popular. Big houses constructed.	Pueblo Grande big house built.
1200		*Soho*	Rectangular platform mounds with compound walls dominate villages. Ballcourt system abandoned.	Platform mound now one mound and rectangular in shape. NW ballcourt abandoned.
1150	**Classic** (1150-1450)		Aboveground residential areas with compound walls. Decline in Hohokam interaction outside Gila-Salt river valleys.	Two small platform mounds built. 1,000 Hohokam living in the village. Formal cemeteries.
1125 1000		*Santan*	Snaketown abandoned. Inhumation burials with polished red wares.	Small caliche-capped mound.
950 900	**Sedentary** (900-1150)	*Sacaton*	Mass production of red-on-buff pottery. Increased use of ballcourts.	Northwest ballcourt constructed.
875 800		*Santa Cruz*	Artistic florescence. Ceramic motifs carefully executed. Elaborate cremation rituals.	Pueblo Grande is a sizable village. Ten irrigation canals in use.
750	**Colonial** (750-900)	*Gila Butte*	First ballcourts. Increased trade in exotic items.	
700		*Snaketown*	First capped low platform mounds. Cremation predominant mortuary practice. Red-on-buff pottery.	Pueblo Grande is a small village.
600		*Sweetwater*	Large irrigation systems on north and south side of Salt River.	Canal System Two developed, with canal heads in the Park of Four Waters.
500 450		*Estrella*	First large irrigation canals. Red-on-gray pottery. Bow and arrow used in the Southwest.	First houses at Pueblo Grande. Pueblo Grande is a small hamlet.
300		*Vahki*	Polished red wares.	
50			First irrigation canals built.	Small amounts of trash deposited.
A.D. 1	**Pioneer** (A.D. 1-750)	*Red Mountain*	Beginning of pottery production. Small permanent villages. Emphasis on farming.	
B.C.	**Late Archaic** (1500 B.C.-A.D. 1)		Hunters and gathers with limited agriculture. Clay figurines. Small pithouses. Seasonally occupied hamlets.	No occupation.

Illustration of typical Sonoran Desert landscape. Courtesy of the Bureau of Land Management.

The Sonoran Desert

Chuckwalla from South Mountains.

To understand the Hohokam, one must understand the desert setting in which they lived, which was a challenging and sometimes difficult world. However, the lengthy adaptation to the Sonoran Desert by the Hohokam and their Archaic ancestors demonstrates that people could make a successful living in this arid yet productive environment. The village of Pueblo Grande is a monument to the accomplishments of the Hohokam people and how they lived in balance with the Sonoran Desert.

The Sonoran Desert is vast, measuring 100,000 square miles (259,000 sq. km), and is one of four major deserts in North America. Other deserts are the Great Basin in Nevada, the Mojave in California, and the Chihuahua in Mexico. Despite its harsh climate and relative scarcity of permanent water sources, the Sonoran Desert of south-central Arizona and northern Mexico is one of the richest deserts of the world in terms of density and diversity of native plants and animals. Data gathered by scientists indicate that the Hohokam lived in a natural environment that was essentially the same as the Sonoran Desert appears today.

Rainfall in the Sonoran Desert is *bimodal* (occurring in both winter and summer) and averages less than 15 inches (38 cm) per year, with some areas such as Gila Bend receiving less than 6 inches (15 cm). In the Salt and Gila river valleys, the heartland of the Hohokam, the historic annual average rainfall is about 7.5 inches (19 cm). Temperatures in the Sonoran Desert vary from below freezing in winter to above 120 degrees Fahrenheit (49°C) during the summer.

Topographically, the Arizona Upland portion of the Sonoran Desert ranges from 700 feet (213 m) to over 3,000 feet (914 m) above sea level. This landscape includes rugged mountains that generally are oriented in a northwest-southeast direction. Rivers that flow out of the mountains cut broad valleys through the hard desert soils. From these river valleys, the desert floor slopes upward and turns into rolling hills which become steeper as they near the mountains. These elevational differences provided for a rich diversity of plants and animals that were available to the Hohokam.

More than 3,500 species of plants grow in the Sonoran Desert, but its vegetation is best known for its tall saguaro cacti, some reaching 50 feet (15 m) in height. Mixed

Saguaro Cacti. The Sonoran Desert is best known for its tall saguaro cactus, the only place in the world these tree-like plants can be found growing naturally. The Hohokam used the inner woody ribs of the saguaro for building material and picked the fruit to make wine as well as for food.

among the saguaro are mesquite, ironwood, and palo verde trees. Vast arrays of other desert plants include creosote, ocotillo, jojoba, cholla, prickly pear, barrel and hedgehog cacti, agave, sotol, and yucca. In the higher elevations of the mountains grow juniper, oak and pine trees.

Hundreds of animal species live in the Sonoran Desert. Reptiles include several species of lizards, tortoises, and a wide variety of nonpoisonous as well as poisonous snakes. Throughout the desert are many types of rodents, including mice, rats, squirrels, jackrabbits, and cottontail rabbits. Muskrat and beaver live in the rivers and streams. Larger mammals include mule deer, white-tailed deer, pronghorn antelope, and bighorn sheep. Predators consist of bears, mountain lions, wolves, coyotes, fox, badgers, and bobcats, as well as hawks and eagles. Birds include quail, doves, and many species of water fowl.

The Sonoran Desert is crossed by several rivers such as the Salt, Gila, Verde, San Pedro and Santa Cruz, which in prehistoric times flowed year round and created lush green tracts of land winding through the desert. These wide ribbons of flowing water supported many types of riverine plants and animals. In prehistoric times, before the major rivers were dammed, there was also an abundant supply of several species of fish.

It was along these rivers that the Hohokam settled down into agricultural villages around the beginning of the Common or Christian era (A.D. 1). One of these villages was the site of Pueblo Grande. Over several centuries, Pueblo Grande would become one of the most important Hohokam villages along the Salt River.

The Pueblo Grande Village

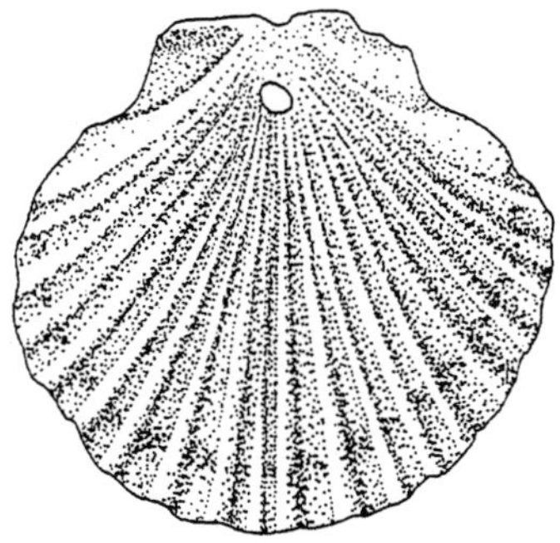

Pecten shell pendent from Pueblo Grande.

The Hohokam village ruin located on the north side of the Salt River and two miles west of the Papago Buttes was given the name *Pueblo Grande*, Spanish for *large town*, by City of Phoenix Engineer Omar Turney in the 1920s. He named Pueblo Grande after its most prominent feature, a platform mound, one of the largest of more than 50 such structures built by the Hohokam in the Salt River Valley. Besides the platform mound, the village site contained two and possibly three ballcourts, several major irrigation canals, trash mounds, houses, courtyards, plazas, *hornos* or ovens, compounds, storage pits, cemeteries, and various other cultural features. The village of Pueblo Grande also once included a *big house* (a large multi-story adobe building), similar to the famous big house at Casa Grande Ruins National Monument near Coolidge, southeast of Phoenix. Fed by water from the canals, large cultivated fields of corn, beans, squash, and cotton stretched out from the village in several directions.

Pueblo Grande appears to have been settled sometime before A.D. 500, perhaps related to an early canal system that the Hohokam built at the southern edge of the site in the area known today as the *Park of Four Waters*. Within a few hundred years, Pueblo Grande had grown into a large, thriving village more than a mile (1.6 km) in diameter. The village was strategically located at the headwaters of a major canal system, called Canal System Two by Omar Turney, and would have held an influential position among the Hohokam communities of the Salt River Valley.

The major features of Pueblo Grande were built over generations of time and show the dynamic growth of the village. Early in the site's history, residential areas consisted of clusters of pithouses and other domestic features. Sometime around A.D. 1000, the northwest ballcourt (the only one at the site excavated by archaeologists) was constructed at Pueblo Grande. A century later, two small mounds existed south of the ballcourt, precursors to the platform mound visible today. Sometime around A.D. 1200, the ballcourt was abandoned and the platform mound, now one mound and rectangular, became the focal point of the village. Houses were now predominately constructed of adobe, built above ground and surrounded by compound walls. The Pueblo Grande big house probably was built sometime in the 1300s when as many as ten irrigation canals were in use in the Park of Four Waters, south of the platform mound.

As with other major villages in the Salt River Valley, the Hohokam abandoned Pueblo Grande sometime in the mid-15th century after decades of population decline. One thousand years of occupation ended as the inhabitants of the village faded into legend, speculation, and scientific inquiry.

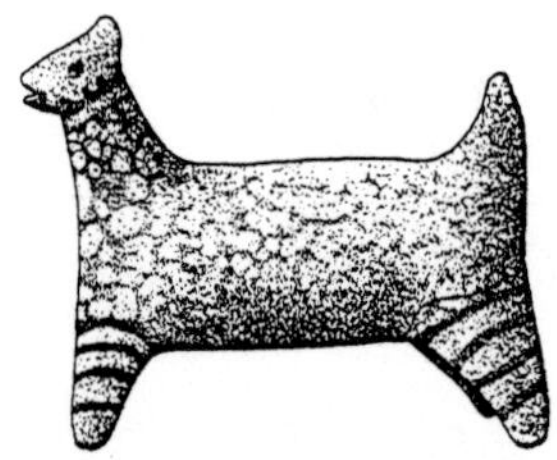

Possible dog figurine from cache found in Habitation Area 5.

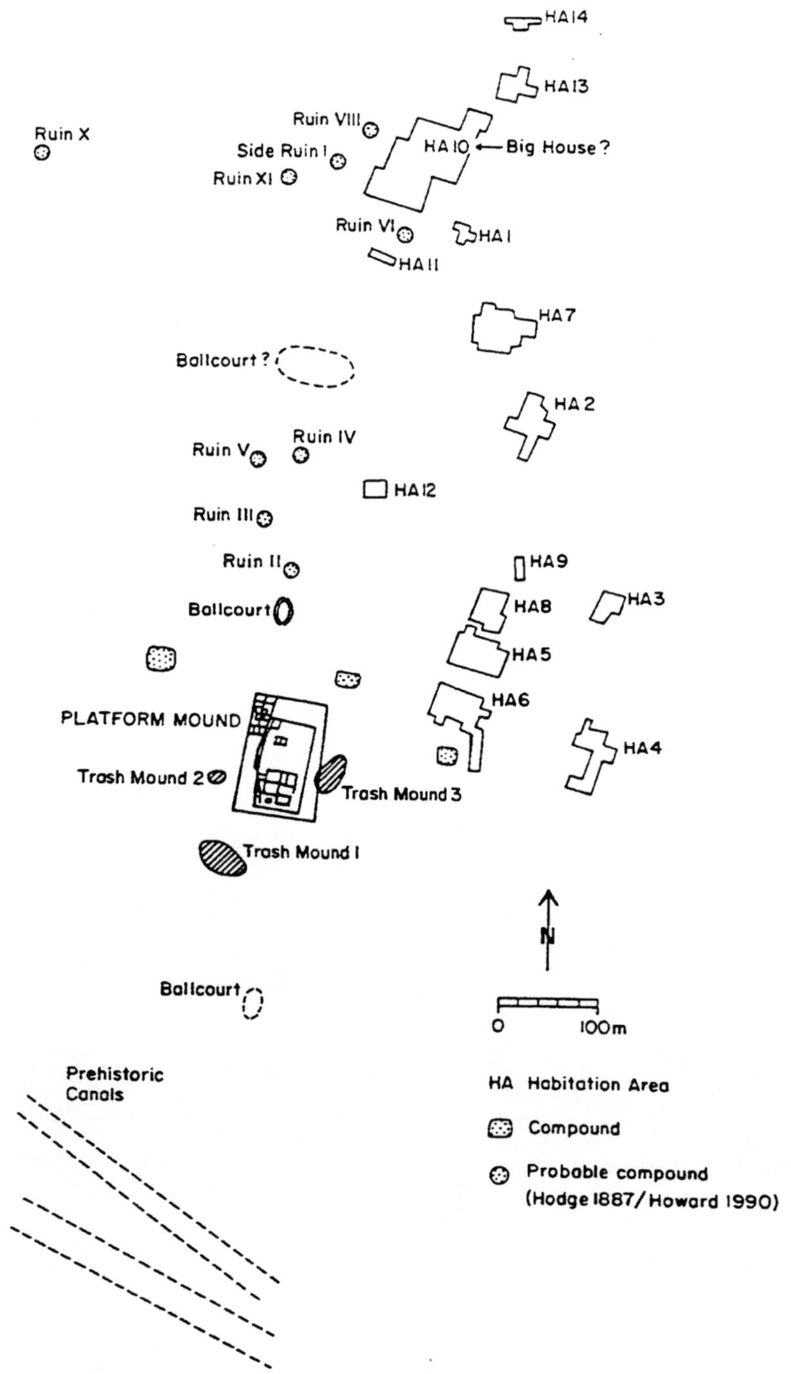

Archaeologist's 1990 map of the Pueblo Grande village showing its most prominent features. The village contained a large platform mound, two or three ballcourts, three large trash mounds, numerous habitation areas and cemeteries, and as many as 20 canals. On this planview, "ruins" are large features recorded by the Hemenway Expedition in 1887, and "HA" are habitation areas excavated by Soil Systems, Inc. for the Arizona Department of Transportation in 1989 and 1990. More habitation areas were discovered by Soil Systems north of the park boundaries in 1999 and 2000.

Harvesting the Desert

Hohokam deer petroglyph from South Mountains.

The Hohokam used various techniques to procure food from the desert environment in which they lived. At Pueblo Grande and other villages, they practiced agriculture using irrigation canals and flood water farming, hunted desert animals and collected local wild plants. From the canals and rivers they obtained fish and other aquatic plants and animals.

Agriculture

In the prehistoric Southwest, the Hohokam were unsurpassed as farmers. With their elaborate system of irrigation canals, they were able to cultivate a large variety of domesticated plants. Corn, domesticated in Mesoamerica around 7,000 years ago appears to have been the primary food of the Hohokam people. Evidence of corn at Pueblo Grande and other Hohokam sites comes from charred kernels and burned corn cobs that have been found in rooms on top of the platform mound and from residential areas.

Other crops grown by the Hohokam included several types of beans (common beans, jackbeans, tepary beans, and lima beans), cushaw and butternut squash, pumpkins, little barley grass, amaranth, gourds, cotton and possibly tobacco. In addition, the Hohokam planted or encouraged the growth of wild desert plants, such as agave (century plant) and cholla cactus, in the drier or marginal areas of their fields.

Although archaeologists do not know how the Hohokam divided their labor among the sexes, *ethnographic* data indicates that for most Southwestern groups, men and women performed different tasks related to subsistence. Men typically hunted game, cleared the land and prepared the soil for gardens. Women planted and harvested the crops and prepared all the food.

Ethnographic data of the historic Pima indicate that agricultural fields were divided into small plots and farmed by extended family groups. Although it cannot be determined through archaeological investigation, the Hohokam, like historic American Indian groups, most likely planted a wide variety of plant species together in their fields. In addition, the Hohokam may have double-cropped, producing two yields per year by taking advantage of two growing seasons.

The Hohokam planted their fields between rows of lateral irrigation canals, as well as along washes and in areas which received adequate runoff from rainwater. Their fields covered thousands of acres in the Salt River Valley alone and produced enough food for tens of thousands of people. Animals attracted to the lush Hohokam fields were actively hunted and likely constituted a major portion of meat in the Hohokam diet. At Pueblo Grande the majority of the agricultural fields were located to the west of the village, among the canals that originated to the south in the Park of Four Waters.

Canals and Irrigation

To farm the desert successfully, the Hohokam tapped the rivers of the Sonoran Desert by constructing irrigation canals. Using simple tools, the Hohokam created the largest prehistoric irrigation system in North America. Many modern canals in the Phoenix metropolitan area are located in the same place as the prehistoric Hohokam canals. Some of the last undisturbed prehistoric Hohokam canals are located at Pueblo Grande in the Park of Four Waters.

The Hohokam experimented with small irrigation ditches early in their occupation of the Salt and Gila river valleys. By A.D. 600, they had developed lengthy networks of canal systems. Eventually, hundreds, perhaps thousands, of miles of these waterways wound out from the Salt and Gila rivers, creating swaths of green vegetation surrounded by sprawling villages. Many of these irrigation canals were over seven miles (11 km) in length. The longest recorded Hohokam canal extended possibly as much as 20 miles (32 km) or more, beginning at Pueblo Grande, and reaching as far west as present day Glendale.

Irrigated corn field in Glendale, AZ. Hohokam fields under irrigation may have looked similar to this modern day irrigated corn field.

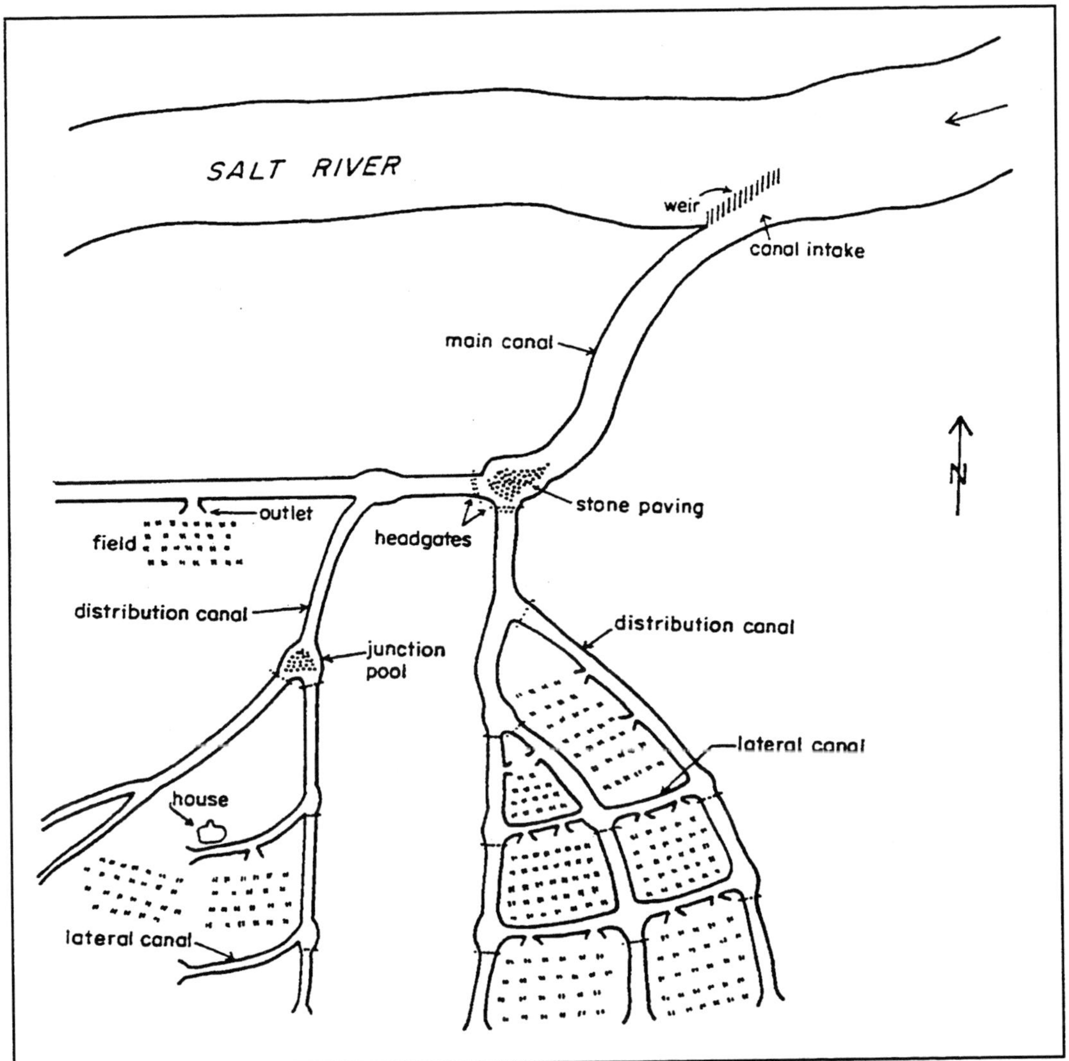

Hohokam canal system components (after Masse in *Chaco & Hohokam*). The Hohokam constructed possibly the largest canal system in the New World. Over a 1,400 year period, the Hohokam dug more than 1,000 miles (1,600 km) of canals, using only stone and wooden tools.

The Hohokam canal system consisted of several integrated parts. To direct water from the river into their canals, archaeologists assume the Hohokam constructed *weirs* at a 45 degree angle to the flow of water. A weir was a type of log and brush dam that extended into the river but not completely across. In this way the weir raised the water level of the river and forced it into the *main* canal. To control and regulate the flow of water inside the main canal, a *headgate* was constructed, also of logs and brush. The main canals carried water away from the river and toward villages. Branching off from the main canals were *distribution* canals, which transported water to the fields. Finally, smaller *lateral* canals released the water directly onto the fields.

Canal construction was not an easy task for the Hohokam. Having no draft animals, the Hohokam had to dig by hand through the hard, sunbaked soil using large wedge-shaped stone hoes, stone axes and

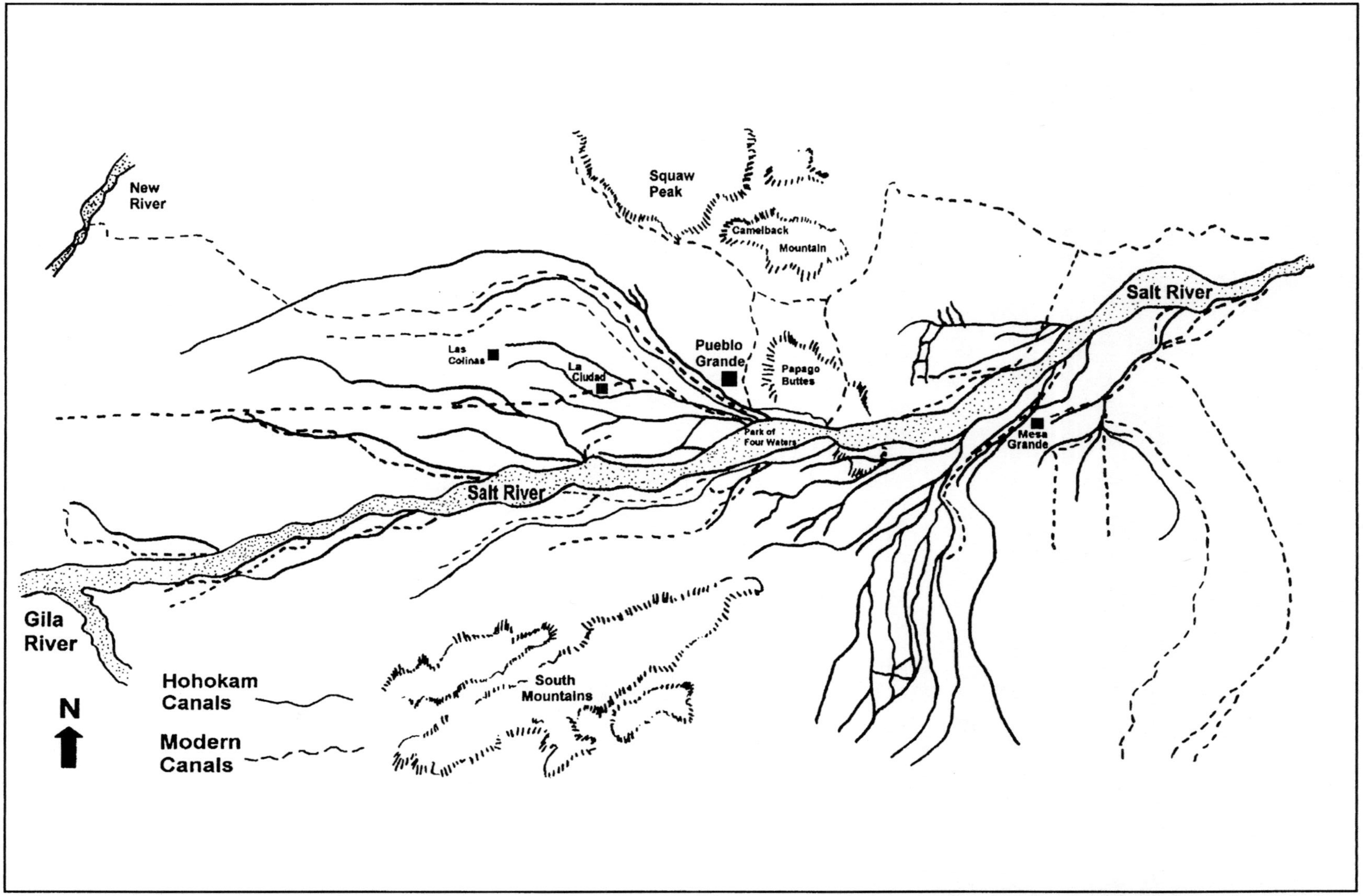

Main canals of the Hohokam irrigation canal system in the Salt River Valley, based on map compiled by Omar Turney in 1929. The Hohokam constructed possibly the most elaborate irrigation system in the Americas. This system contained at least a thousand miles (1,600 km) of canals in the Salt River Valley alone and irrigated over 100,000 acres (40,469 ha) of farmland. Many of the modern day canals used in the Phoenix metropolitan area today are based on this Hohokam system.

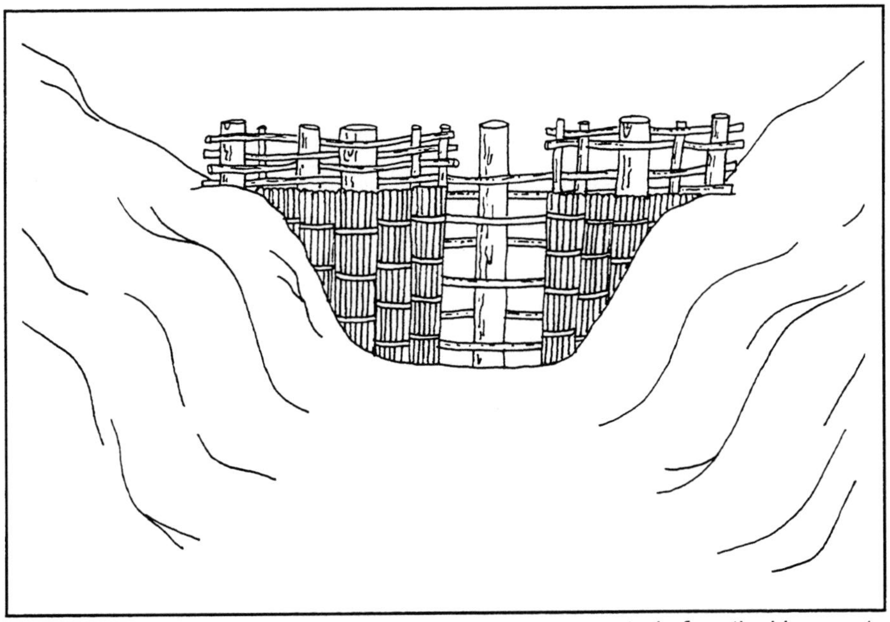

A Hohokam canal headgate. Headgates were constructed of vertical log posts set into the canal bed. Brush and bundles of reeds were interwoven into the posts to create a somewhat watertight barrier. Like headgates in modern day canals, these too were used to control the speed and amount of water flow.

digging sticks. The excavated dirt probably was then removed in baskets and used to build embankments along the sides of the canals. Some canals were very large, reaching widths of over 50 feet (15 m). Main canals at Pueblo Grande (Canal System Two) are as much as 30 feet (9 m) wide and 10 feet (3 m) deep, with banks up to 10 feet (3 m) high, making the canals 20 feet (6 m) deep from top to bottom.

Besides the actual digging of the canal, the Hohokam also had to determine the most opportune location for the canal. This required a knowledge of hydraulics. Due to the complexity of the canal system, the Hohokam most likely would have had to lay out the course of the canal from start to finish before construction could begin. In locating and building their canals, the Hohokam may have used simple surveying equipment to determine elevations and gradients over long distances.

Hohokam canals were engineered in such a way that they were wide at the headwaters and gradually narrowed as they reached their *terminus* or end. In this way, water levels were kept constant and there was an even flow of water from beginning to end. Canal gradient or slope was crucial for the proper working function of the canals. If the grade was too steep, the water would run too fast and erode the canal banks. On the other hand, if the grade was too gentle, the water would run slow, dropping its sediment load and causing the canal to silt up. Either one of these malfunctions would have required countless hours of labor to remedy.

Hohokam canals were constructed with different gradients, depending on the canal's location within the system and its distance from the river. Canal gradients within Canal System Two, for example, varied from 4.8 feet (1.5 m) per mile (1.6 km), to 1.2 feet (0.4 m) per mile (1.6 km).

The topography of the land surrounding Pueblo Grande made it an ideal location for the headgates of irrigation canals. The site is located just west of Tempe Butte, where the bedrock is close to the surface. The river bends to the south at Pueblo Grande, making it a favorable location for irrigation canals heading in a northwesterly direction. In addition, the contour of the land in this area is relatively flat and uniform, allowing for several canals to diverge from the river for considerable distances. During the Classic period (A.D. 1150-1450), as many as ten canals were in use at the same time at Pueblo Grande, but many of those canals carried water to fields at other villages.

The Hohokam canal system was built over several centuries, and not all Hohokam canals identified by archaeologists were in use at any one time. Over a period of more than a thousand years canals were built and abandoned as water flow changed, headgates were destroyed by floods, or canals were filled by sediment deposition.

According to archaeologist Jerry Howard, the greatest amount of canal construction in Canal System Two occurred early in the Hohokam chronological sequence during the Colonial period (A.D. 750-900). During this period, canals damaged by floods were regularly rebuilt. By the beginning of the Sedentary period (A.D. 900-1150), canals closest to the Salt River were abandoned and new replacement canals were constructed on higher ground. During the Sedentary and Classic periods, the Hohokam consecutively built canals further and higher from the river while discontinuing use of some of the older southern canals.

Abandonment of canals also led to the abandonment and shifting of villages from the south (near the river) to the north (away from the river). An exception to this trend was the construction during the Classic period of a new, independent canal system and village at Pueblo Salado, just north of the Salt River where the Sky Harbor Airport is now located. This probably was a reclamation effort to farm land previously abandoned during the earlier Colonial period.

The Hohokam canal system and all of its elaborate parts indicate that the Hohokam had a somewhat complex sociopolitical structure. Beyond the canals themselves, this complexity is expressed during the Classic period in the location of villages containing earthen platform mounds approximately every three miles (5 km) along the irrigation canals. This suggests that the theocratic or secular elite of these platform mound villages controlled the construction of the canals plus the flow of water in them, their maintenance, and water allocation.

At Pueblo Grande, it is estimated that the canal system originating on the southern edge of the site irrigated at least 10,000 acres (4,047 ha) of farmland. Pueblo Grande is situated at the headwaters of this canal system that controlled the water for the majority of the Hohokam villages on the north side of the Salt River. Because of its strategic location, Pueblo Grande would have been very important to other villages using the canals along the same canal system. This may be the reason why the platform mound at Pueblo Grande, containing as much as 50,000 cubic feet (1,416 cubic meters) of earth and stone, is the largest of all the platform mounds located along Canal System Two.

Hohokam rock art hunt scene in the South Mountains. Several petroglyph panels in Hohokam territory show humans hunting deer and mountain sheep with bows and arrows. The bow and arrow were probably not introduced into the Southwest until after A.D. 500.

Cottontail Rabbit. Cottontail rabbits and jackrabbits were the most frequently eaten animal by the Hohokam. These animals were probably found in abundance near or in their fields.

Hunting and Gathering

The Hohokam harvested a long list of wild foods. Bean pods were obtained from mesquite trees (and also from other desert trees). Desert shrubs provided seeds and fruit. Buds and fruit from cacti such as the saguaro, cholla and prickly pear were eaten. Acorns were gathered in the mountainous areas, hearts of agave plants were roasted, and various roots, bulbs, and herbs were collected to add to their diet. The Hohokam took advantage of the fact that there are more than 200 species of edible wild plants in the Sonoran Desert. In some areas, the Hohokam may have set fire to tracts of land to promote the growth of certain grasses and cool season herbs they favored.

The Hohokam had only one domesticated animal, the dog, and there is no evidence they ate them. However, they did eat a wide range of wild desert animals. To accomplish this, the Hohokam probably employed a number of hunting techniques and devices. One of these, the bow and arrow, was adopted by cultures of the Southwest as early as A.D. 500. Although no actual bows have been recovered from Hohokam sites, it is known that the bow and arrow were used based on figures painted on pottery, images pecked in stone as *petroglyphs*, and from the recovery of stone *projectile points* or arrowheads. Another hunting device, known as the *atlatl* or spear-thrower, was used by earlier cultures of the Southwest and may have been used by the Hohokam as well. The Hohokam also are believed to have used traps, snares and nets to obtain fresh meat.

Based on archaeological evidence from trash deposits at Pueblo Grande and other Hohokam sites, the list of animals the Hohokam ate includes various species of rodents, birds, jackrabbits, cottontail rabbits, mule deer, white-tailed deer, pronghorn antelope, and bighorn sheep. Other food sources consisted of desert tortoise, mud turtles, and several species of water fowl. Rabbits appear to have been the main source of animal protein for the Hohokam and would have been readily available not far from their villages, as well as in or near their fields.

Ethnographic data show that for most Southwestern groups small game was hunted by men, women, and children, often while gathering wild plants or tending fields. Large game, however, was hunted by small parties of men and involved certain rituals before and after the hunt. The Hohokam may have followed similar hunting procedures.

Domesticated dogs are often found buried at Hohokam villages, and canines may have served as pets and were used for hunting. More than a dozen dog burials have been excavated from Pueblo Grande, and the presence of bones of young dogs suggests that the Hohokam may have raised dogs at the site.

Fishing

Archaeological evidence indicates that the Hohokam fished the major rivers and streams of the Sonoran Desert. Many of the larger rivers ran year round and would have provided abundant fish as a source of protein. Hohokam canals also would have contained fish that could have been caught using fishing nets and traps. Species of fish found at Pueblo Grande and other Hohokam sites include minnows, chubs, Colorado River squawfish, and suckers. Gila Mountain razorback suckers can achieve weights of up to 30 pounds (13.6 kg) and would have provided the Hohokam a hearty meal.

Also living in the rivers and canals was a large molluscan community, including California floater clams, fingernail clams, and assorted snails. The presence of these clams and snails at Hohokam sites suggests they were an important supplemental protein source.

It is not clear, however, how important fish were to the Hohokam diet. Most excavations have not used screens with small enough mesh to recover fish bones. Although by using fine screens, archaeologists have recovered thousands of fish bones from trash deposits at Pueblo Grande. These deposits have been dated to the Classic period (ca. A.D. 1150-1450). It is possible that the Hohokam had increased their exploitation of riverine resources due to an inadequate amount of protein in their diet during that time period.

Conclusions

At first appearance, the Sonoran Desert looks like a harsh and forbidding environment. However, it is much more luxuriant and diversified than recognized by many modern people. Although arid, the Sonoran Desert provides a diverse range of useful plant and animal products. The Hohokam developed several very successful strategies for living in this environment and, therefore, were able to achieve many cultural accomplishments. Some of those achievements, described in the following pages of this book, include domestic architecture, craft goods, pottery vessels, jewelry, cotton clothing, ballcourts, platform mounds, and the establishment of a rich and widespread trading network.

Desert Farmers at the River's Edge. Artist's reconstruction of a field house at the edge of a Hohokam village, with irrigation canals and fields in the background.

Of Dirt and Wood

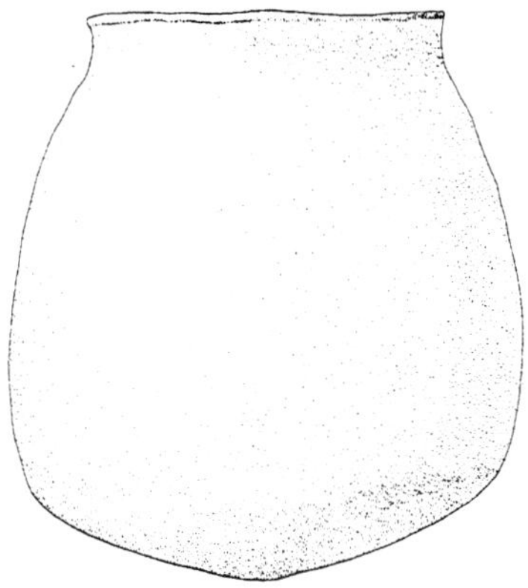

Large plainware vessel used for storage from Pueblo Grande.

The Hohokam constructed a number of features used for domestic purposes. This type of architecture would have been constructed and used by immediate or extended family groups. Domestic architecture among the Hohokam included pithouses, coursed-adobe houses, field houses, hornos, trash mounds, storage pits, and ramadas.

Pithouses

The most common type of housing built by the Hohokam were pithouses. This type of house is distinct by its placement in a shallow pit dug about 1.5 feet (0.5 m) deep. The house was constructed inside the pit and consisted of a wooden superstructure built of mesquite or cottonwood beams interlaced with sticks, saguaro ribs, cholla branches, and grasses. These were then covered with mud and adobe. Burying the pithouse walls kept out drafts, thus insulating the house against the wide extremes of desert temperatures. These houses would have kept occupants relatively cool in summer and warm in winter.

Hohokam pithouses were predominantly single-roomed dwellings that were oval or semi-rectangular in shape and had small, sheltered entrances which protruded from one side of the structure. Just inside

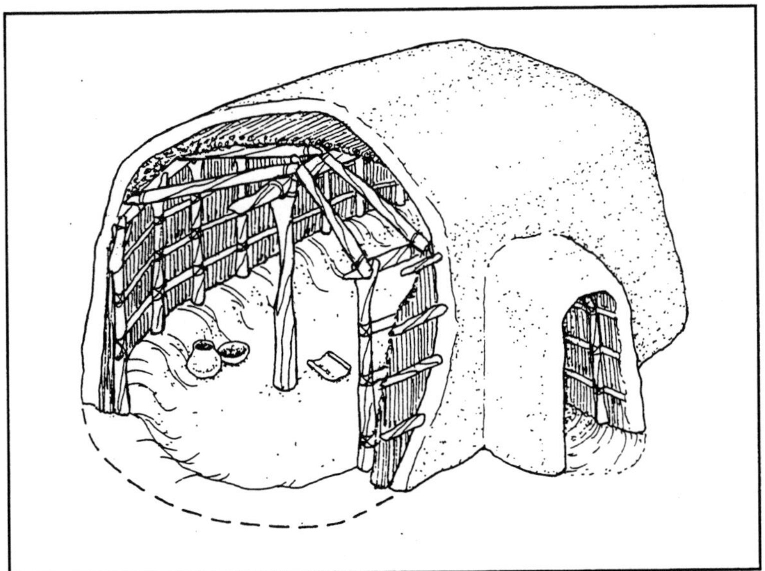

Cutaway view of a Hohokam pithouse. Throughout much of their history, Hohokam houses were built in shallow pits about 1½ feet (.5 m) deep. These pithouses, with their thick mud walls over a wooden superstructure, offered the Hohokam protection against the wide range of desert temperatures.

the doorway a small hearth was built that would have contained hot coals to be used for heat during the winter and perhaps for some light, since these houses are believed to have contained no windows. Most activities then were probably done outside under open-sided ramadas placed near the house. Inside some Hohokam pithouses there is evidence for the existence of raised areas, possibly used for sleeping. In addition, storage pits are often found in Hohokam pithouses, although these features are also common outside. Cooking was done outside in small roasting pits or in semi-subterranean ovens called *hornos*.

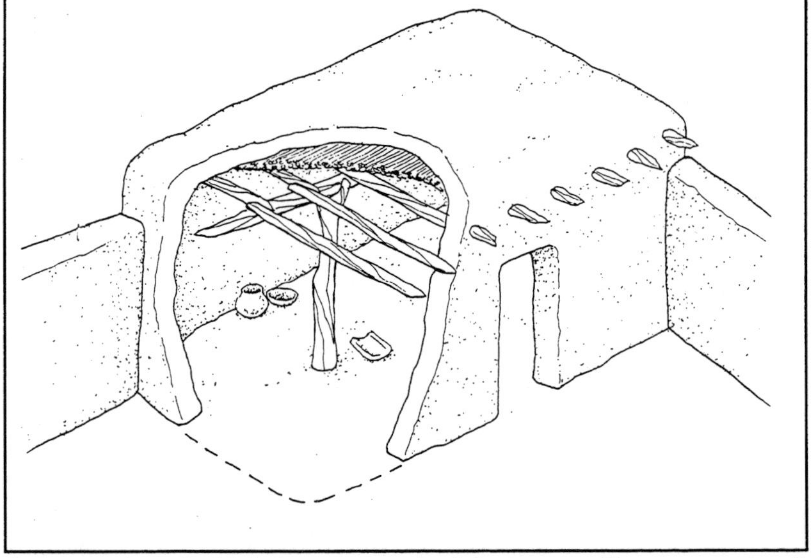

Cutaway view of a Hohokam coursed-adobe house with compound wall. This type of house was built by the Hohokam during the Classic period. The walls were built of adobe mud stacked in layers or courses. The roofs were made of wooden beams covered in adobe in much the same way as the earlier pithouses. Houses during this period were built adjacent to each other in clusters and surrounded by a compound wall.

Pithouses at Pueblo Grande and other Hohokam sites were built in clusters and many of the structures faced inward toward a common courtyard, or open activity area. These clusters probably represented kin groups or extended (related) families. More than 150 pithouses have been excavated at Pueblo Grande, though this number is believed to be only a fraction of the total number of house structures built over the course of ten centuries. By studying the spatial relationship of these houses, archaeologists have been able to determine site growth through time, as well as the internal structure of the Pueblo Grande village. The earliest house built at Pueblo Grande dates to the Pioneer period, ca. A.D. 450. Only a few houses were occupied at the site during this period, however, and it was not until the following Colonial period (ca. A.D. 750-900) that a village existed at Pueblo Grande.

Coursed-Adobe Houses

Although the Hohokam always built pithouses, during the Classic period (A.D. 1150-1450) aboveground houses became much more prevalent. Instead of being constructed of wood and mud, these square or rectangular houses were built of *coursed-adobe*, which was essentially mud stacked in layers with stone sometimes used at the base of the wall. As with pithouses, the roofs of these structures were still constructed of wooden beams interlaced with sticks, saguaro ribs, and grasses. Mud was placed over these materials creating a roof that also served as a floor for some multi-storied structures.

Most Hohokam aboveground structures were built adjacent to each other in clusters, as were the earlier pithouses. However, the aboveground structures were surrounded by an adobe wall, creating an apartment-like complex. As many as 20

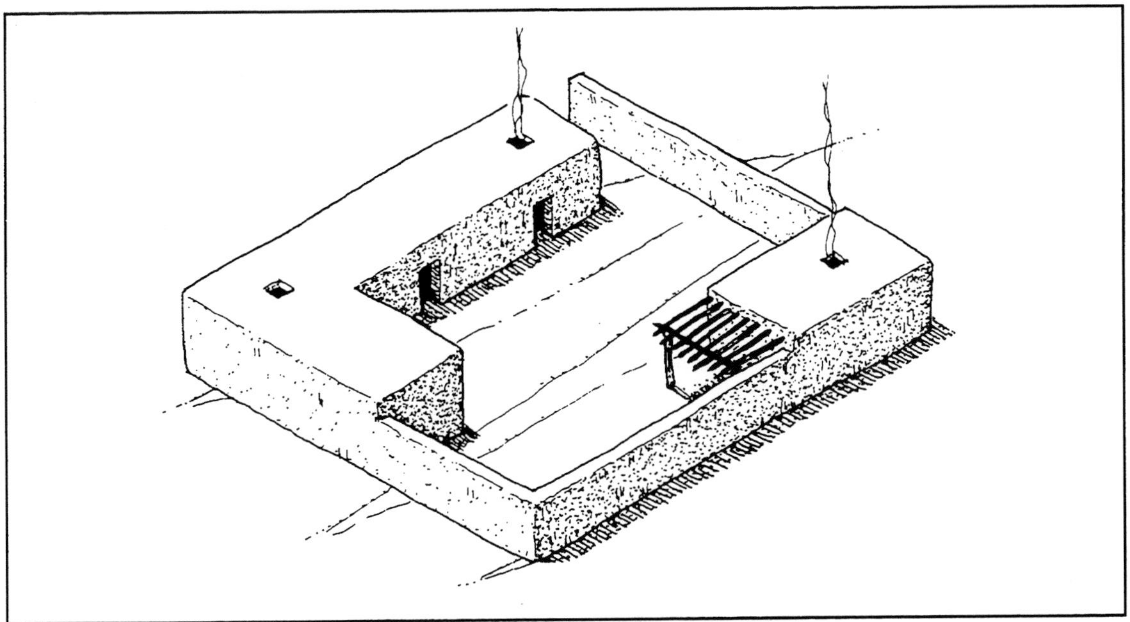

Artist's reconstruction of a typical Hohokam residential compound. During the Classic period residential compounds began to be constructed by the Hohokam. These compounds consisted of several rooms, ramadas, and outdoor work areas. An extended family group of several related people may have lived in each compound.

domestic compounds have been found at Pueblo Grande, most of them located north and east of the massive platform mound. These compounds contained up to 17 rooms each and may have housed as many as 35 people. Together, the residential compounds at Pueblo Grande were home to about 1,000 Hohokam.

Field Houses

In or near their agricultural fields, the Hohokam built small field houses. These round or oval shaped brush structures were seasonally occupied and served as temporary residence for field workers performing various tasks. In addition, they may have been used when certain types of work were being done such as canal construction and maintenance. Unlike pithouses and *coursed-adobe* houses, field houses do not appear to have been built in clusters but were scattered among the fields. Extended family groups may have had more than one field house located in different field areas, each relating to separate plots of land.

Hornos

The Hohokam prepared much of their food in family or communal roasting pits or *hornos*, which is Spanish for oven. Hohokam hornos were cone-shaped, semi-subterranean, stone-lined features in which food was baked. A fire was built in the bottom of the pit and then smothered. A layer of stones was then placed on top of the smoldering fire. The food to be cooked was placed on top of the heated stones between layers of grass and dirt, and another fire was built on top. In this way the horno acted like an oven, heating the food from the top and bottom.

Remains of agave hearts and cholla buds are commonly found in hornos ex-

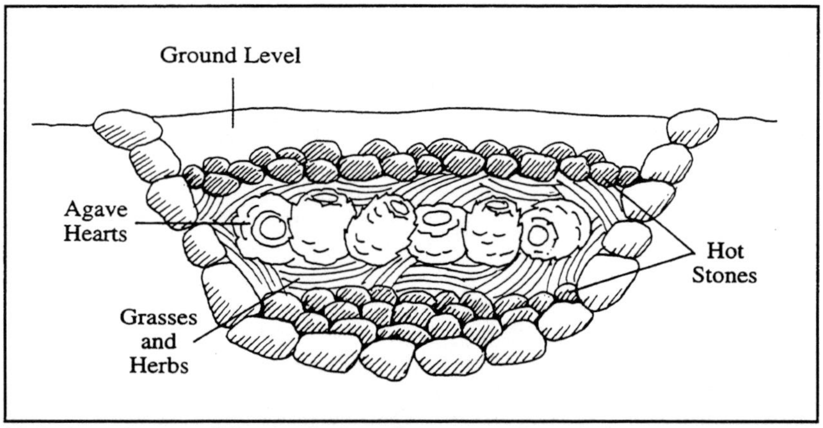

Hohokam horno. Often called an agave roasting pit, hornos were semi-subterranean ovens used by the Hohokam to bake agave hearts, cholla buds and other plant materials. (Illustration adapted from Desert Botanical Garden graphic.)

cavated by archaeologists, suggesting that the Hohokam baked those plants. Interestingly, animal bones are usually scarce or altogether lacking in hornos. Instead, meat was probably roasted over open fires or cooked in pots in small quantities.

It has been suggested that Hohokam hornos may have been used for more than just preparing food for a few families. Some hornos are quite large, measuring as much as 10 feet (3 m) in diameter and almost 8 feet (2.5 m) in depth. The large quantity of food cooked in these ovens could have been used in ceremonial feasts associated with ritual events that took place within a village on special occasions.

Trash Mounds

Trash mounds, although not formal domestic architecture, are the direct result of domestic activities. Like all societies throughout history, the Hohokam had to dispose of their refuse. Typically, they deposited their trash in certain locations adjacent to residential precincts within their villages. These disposal areas were sometimes used for hundreds of years. Over time, trash deposits accumulated into large circular shaped mounds, some reaching heights of 6 feet (2 m) or more. A trash mound excavated by Albert Schroeder at Pueblo Grande in the late 1930s was 118 feet (36 m) wide and 164 feet (50 m) long, with trash deposits recovered to a depth of 8 feet (2.5 m).

Hohokam trash mounds provide a sort of time capsule which reflects the occupational history of a village. Since trash was deposited in layers, excavation of those layers, if undisturbed, reveals a chronology of change through time in material culture at the site. Thus, a temporal sequence can be established, with the earliest trash located at the bottom of the mound and the latest trash at the top of the mound. Changes in ceramic styles can be determined from an analysis of the trash, which is partially composed of large quantities of broken pottery vessels, called *sherds* by archaeologists.

To control pests and odor emanating from the mounds, the Hohokam appear to have occasionally burned debris and capped the mounds with clean desert soil. Because of this practice, perishable materials such as food remains are often preserved in the trash mounds. Emil Haury, one of the pioneers of Hohokam archaeology, proposed that the capping of trash mounds at the site of Snaketown, located next to the Gila River, eventually led to the development of platform mounds.

At Pueblo Grande, three large trash mounds and many smaller ones were built up over centuries of occupation at the village. In 1925, Erich Schmidt of the

American Museum of Natural History excavated one of those trash mounds (Trash Mound No. 2), located under the present Pueblo Grande Museum building. Schmidt's work revealed a long occupational sequence, which he used for his doctoral dissertation at Columbia University, the first one on Hohokam prehistory. An interesting find by Schmidt was a copper bell. This copper bell was the only one discovered at Pueblo Grande until 1999, when eight copper bells were excavated about one half mile to the north of the platform mound by Soil Systems, Inc. Copper bells are rare items in Hohokam sites and were imported from Mexico, suggesting trade relations between the Hohokam and cultures to the south.

Archaeologist's trench in Hohokam trash mound at Villa Buena, ca. 1939. Person on the left is shoveling dirt into a screen that is being shaken by person on the right. Steps or levels cut into the trash mound allows for the study of stratigraphy and the establishment of a temporal sequence based upon the artifact assemblages encountered at each level.

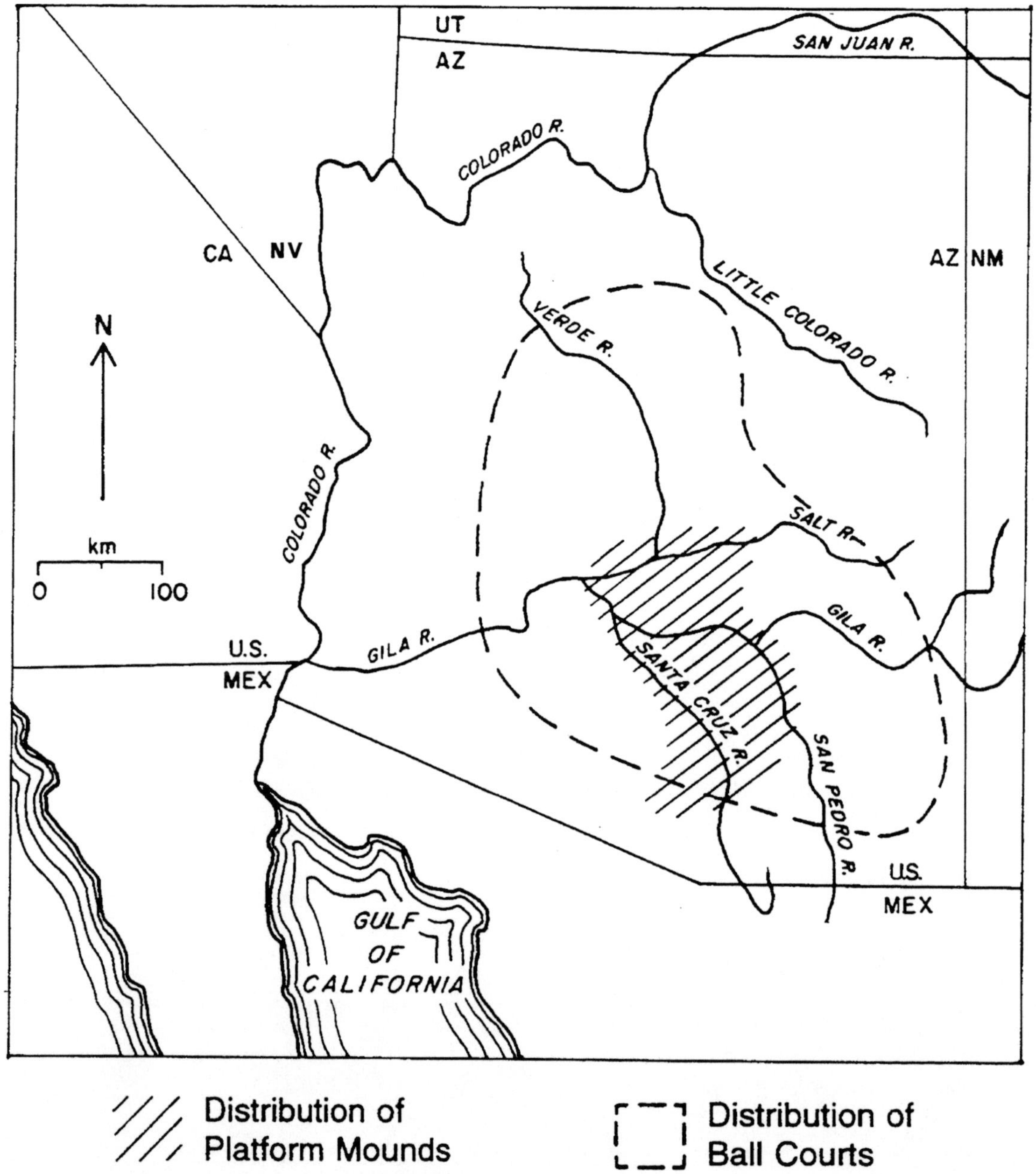

Distribution of Hohokam ballcourts and platform mounds. Ballcourts had a much wider distribution than the later platform mounds, coinciding with the Classic period decline of Hohokam interaction outside the Salt and Gila river valleys and the Tucson Basin.

Ritual, Ceremony, and the Elite

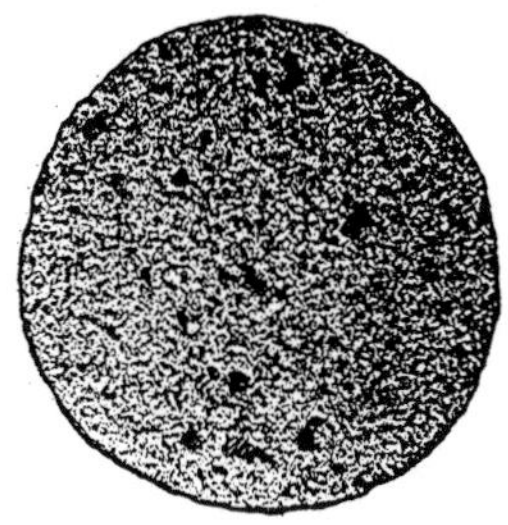

Stone ball from Pueblo Grande ballcourt.

The Hohokam built a variety of non-domestic architectural monuments and structures that required the combined labor of a multitude of people. This type of architecture is called monumental or public architecture and was created for specialized use. Hohokam monumental architecture appears to have been very important in Hohokam society and included ballcourts, platform mounds, and big houses. It was not only the structures themselves that were important to the Hohokam, but the activities associated with them and the broader ideological meaning of those activities.

Ballcourts

One of the more interesting features of Hohokam culture is the ballcourt. More than 200 ballcourts have been recorded in the Hohokam cultural region, although only a few have been excavated. In the Salt River Valley, more than 30 ballcourts from more than 20 villages have been identified. These ballcourts were built and used for nearly 500 years, from about A.D. 750 to 1200, when the system was abandoned.

Hohokam ballcourts are oval, bowl-shaped depressions that vary in size but average 80 to 115 feet (24.5 to 35 m) in length and 50 feet (15 m) in width. Berms of earth, up to 9 feet (3 m) high, encircled the depression and were constructed using the excavated dirt. During a Hohokam ball game, dozens and perhaps hundreds of people could watch from the embankments surrounding the ballcourt. The largest Hohokam ballcourt, excavated in the mid-1930s at Snaketown, 20 miles (32 km) south of Pueblo Grande, was large enough for 500 people to line its banks.

The playing surface of the ballcourts was smoothed and plastered with caliche, producing a symmetrical, concave floor surface that sloped slightly up toward the edges and then rose steeply onto the embankments. At each end of the ballcourt, constricted openings were constructed which may have been goals into which players attempted to place rubber or stone balls. In the northwestern ballcourt at Pueblo Grande, stone markers were embedded in the floor in front of these possible goals, as well as in the middle of the court, possibly denoting zones of play.

Ballcourts apparently were not restricted space, but appear to have been open to the general public. They may have served to promote market activities or ceremonial exchange systems, both within the Hohokam culture and between the Hohokam and other groups.

The exact nature of the games played in these courts, however, is unknown. Some archaeologists argue that they may have been similar to games played in Mesoamerica, and ballcourts and trade items found in Hohokam sites suggest a significant connection with cultures from Mexico and further south. For Mesoamerican peoples, ballcourts were symbolic of the passageway between the upper and lower spiritual worlds. Humans could communicate with the gods through playing the ball game, which was analogous to a mythical drama, expounding world creation myths and beliefs.

Ballcourts in Mesoamerica, however, are physically different from Hohokam ballcourts. Courts in Mesoamerica have flat playing surfaces, are made of stone, and have walls at right angles to the floor. In addition, some Mesoamerican ballcourts contain stone rings in the sides of their walls. It was through these rings that players attempted to place rubber balls. No evidence of similar rings has been found in Hohokam ballcourts.

Unfortunately no ethnographic accounts exist of the Hohokam playing a ball game in ballcourts since they were no longer in use when Spanish explorers arrived in Arizona in the mid-1500s. However, Hohokam ceramic figurines of individuals wearing pads on their arms and hips, simi-

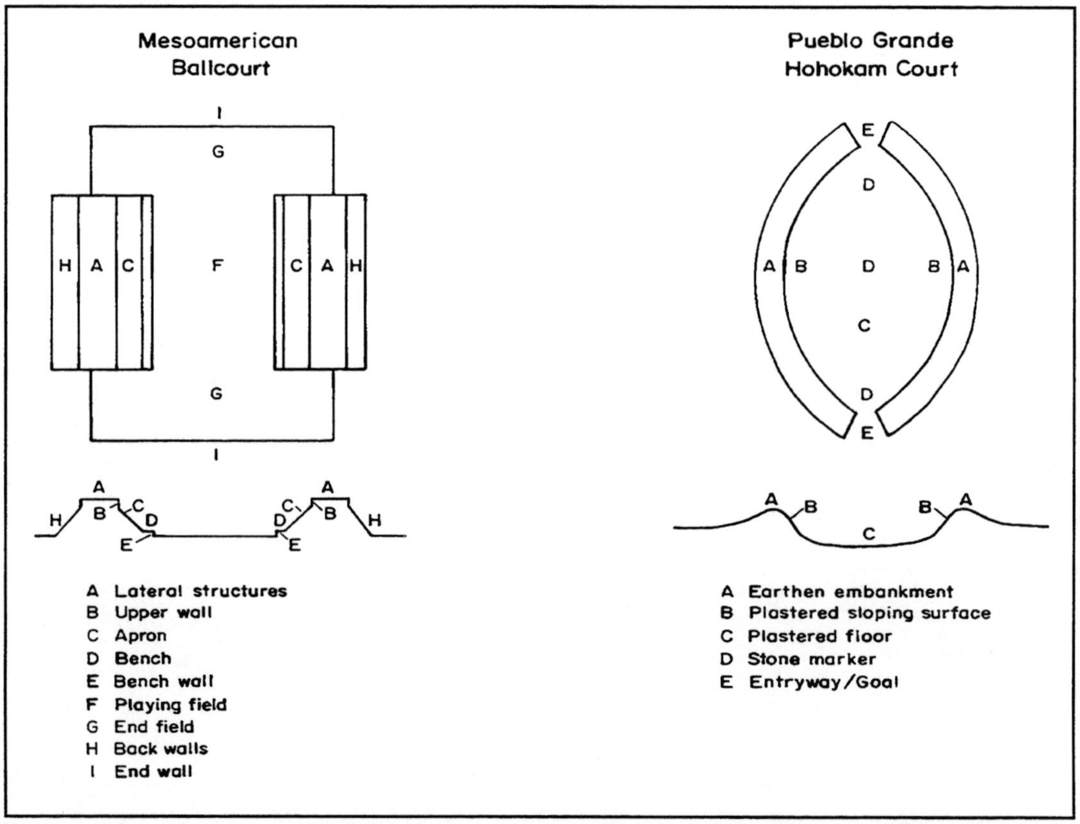

Comparisons of a Mesoamerican ballcourt and the northwest Pueblo Grande ballcourt. Although archaeologists may never know the type(s) of game(s) played in Hohokam ballcourts, there are interesting similarities in the architectural features of Mesoamerican and Hohokam ballcourts.

Artist's reconstruction of Hohokam ball players in the northwest ballcourt at Pueblo Grande. The Hohokam may have played a type of kick ball like the one shown here, or one in which players could only hit the ball with their hips, similar to ball games in Mesoamerica. Hohokam balls may have been made of stone and left uncovered or balls may have been covered with leather or pitch.

lar to Mesoamerican ball players, have been found, as have petroglyphs of possible Hohokam ballplayers. In addition, stone and rubber-like balls that may have been used in ball games have been discovered at some Hohokam sites.

At Pueblo Grande, two and possibly three ballcourts were constructed. Few Hohokam sites have more than two recorded ballcourts (the village of Villa Buena had five). One of the ballcourts at Pueblo Grande, located about 650 feet (198 m) northwest of the platform mound, was excavated in the 1950s. This relatively small ballcourt measures 82 feet (25 m) long and 38 feet (11.5 m) wide and is oriented north-south. It was constructed ca. A.D. 950, late when compared with the construction of many other Hohokam ballcourts. The court was large enough to accommodate only two to four players on the floor at any one time.

A stone ball (see page 25) was excavated from the fill of this court, although it is not known if this ball was one that was actually used in play at the ballcourt, or if it was deposited there after the court was abandoned. This is the only stone ball thus far excavated from a Hohokam ballcourt, although similar stone balls have been recovered from other cultural features (i.e., trash mounds) at Pueblo Grande and other Hohokam sites. Accounts of Mesoamerican ball games suggest the predominant use of rubber balls during play, and a prehistoric rubber-like ball has been found at a Hohokam site in southern Arizona. This rubber-like ball

Northwest Ballcourt at Pueblo Grande looking south toward the platform mound and museum. At least two and possibly three ballcourts were built at Pueblo Grande. Similar to other Hohokam courts, this ballcourt is an oval, bowl-shaped depression with a caliche plastered floor and goal-like openings at each end. Like a modern-day sports arena, a high berm surrounds the ballcourt, acting as seats for spectators to watch the ceremonial games being played.

was made from the resin of the native Sonoran Desert guayule plant.

Sometime after A.D. 1200, the Hohokam of Pueblo Grande discontinued their use of the northwest ballcourt and began using it as a trash deposit. The reason for its abandonment is unknown, but it coincides with a general abandonment of ballcourts throughout the Hohokam cultural region, which may represent a change in Hohokam ideology. It was during this time, the start of the Classic period, that major changes occurred within the Hohokam culture.

Platform Mounds

During the Classic period (A.D. 1150-1450), platform mounds dominated major Hohokam villages, including Pueblo Grande. In the Salt River Valley alone, over 50 platform mounds at more than 30 villages have been identified by archaeologists. Today only a few of those mounds remain. Thanks to a few early preservation-minded individuals, the platform mound at Pueblo Grande is one of those preserved.

Hohokam platform mounds appear to

Southwest corner of the Pueblo Grande platform mound. The platform mound was completely artificial and was constructed beginning about 800 years ago. The outside retaining wall, seen here, was built of caliche rocks, quarried granite, and sandstone slabs, held together by adobe mortar.

destroyed and the architectural remains were used as fill for the new construction.

Platform mounds were surrounded by a high compound wall which restricted access to the mound. Structures for residential or ceremonial use were built on top of the platform mound and inside the compound wall. Some archaeologists have suggested that the elite of Hohokam society lived on top of the platform mounds, while others argue that the mounds served as ceremonial temples where important rituals were performed to insure sufficient rainfall and agricultural success.

Researchers have hypothesized that Hohokam platform mounds were tied to the organization and operation of the canal systems. The largest platform mounds occurred at the heads of major canal systems, such as at Pueblo Grande. Being in control of a vital desert resource such as water most likely enhanced the socio-political and ceremonial importance of the platform mound. The regular spacing of

have begun as low, circular mounds during the 800s, but by the 12th century they had evolved into large rectangular mounds. Classic period platform mounds were constructed and later expanded by building adobe and masonry-walled cells. These cells were filled with trash and soil and then capped with a layer of *caliche* plaster made from a natural calcium carbonate material found in desert soils. During remodeling and expansion, structures built on top of the caliche-capped cells were

Artist's reconstruction of Hohokam workers building a platform mound. Most likely all members of society helped in the construction of these mounds. Small basins were dug for the mixing of adobe, to be used as mortar and plaster. At the Pueblo Grande platform mound, some rocks may have been carried from sources several miles away.

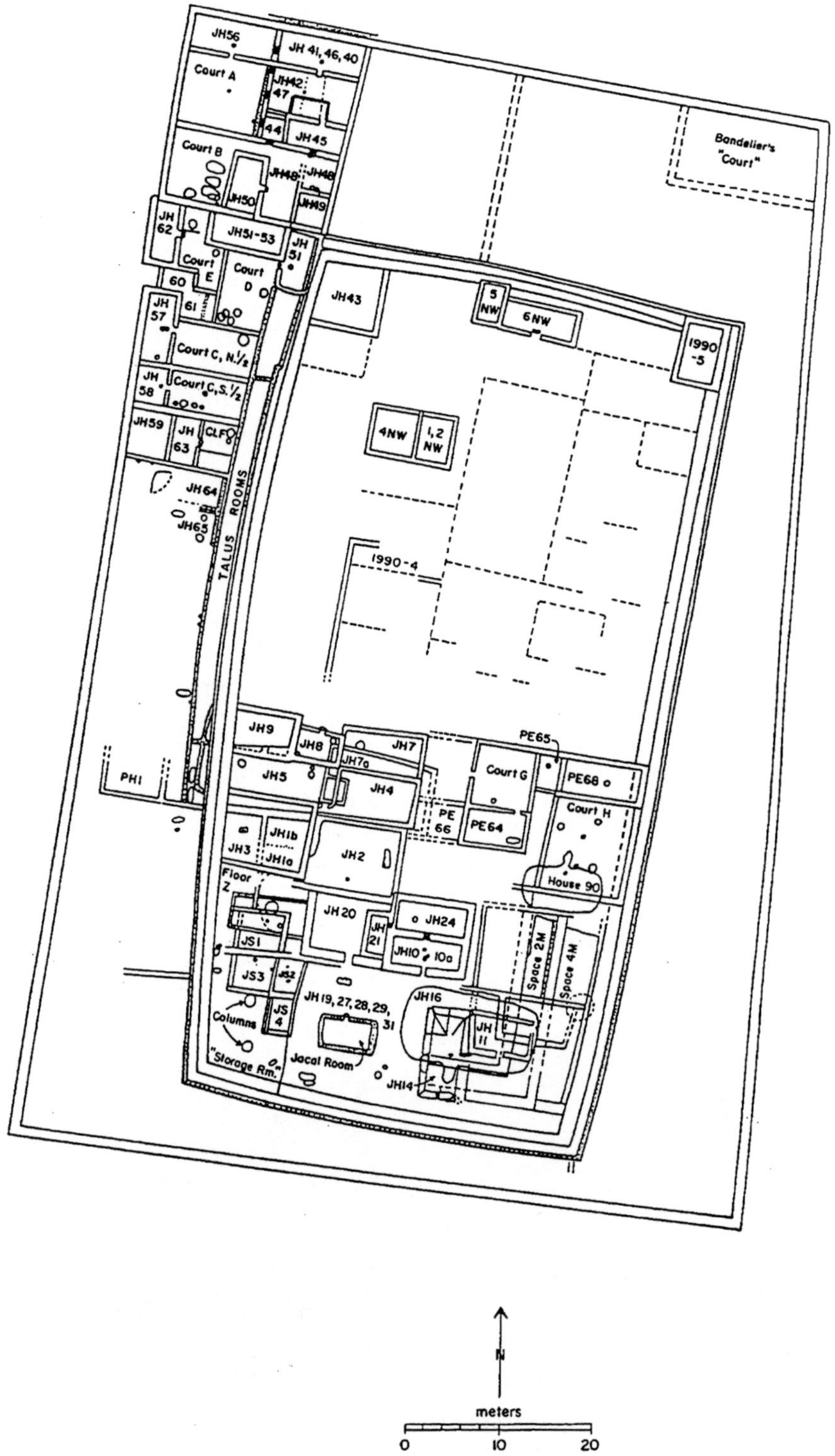

Planview of the Pueblo Grande platform mound (data compiled by Christian Downum). The mound itself, as well as rooms on top of the mound, underwent numerous changes in size and configuration during its three centuries of construction, use and remodeling. Many of the rooms on top of the northern half of the platform mound have not been excavated. Archaeologists have obtained only limited knowledge of this area.

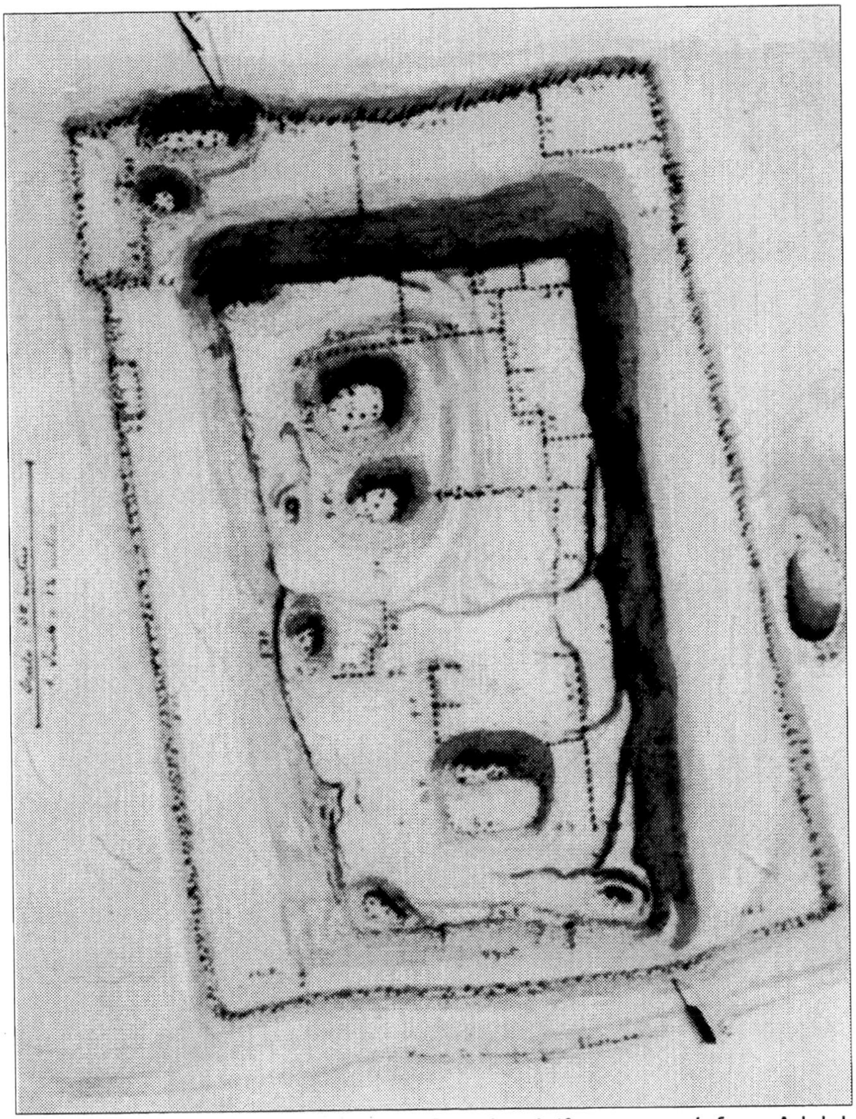

Watercolor painting of the Pueblo Grande platform mound, from Adolph Bandelier's visit in 1883. Bandelier's watercolor is remarkably accurate and has been invaluable for archaeologist's attempts to reconstruct the mound's overall design and layout. (Courtesy of the Vatican.)

platform mound villages, approximately every three miles (5 km) along the irrigation canals, suggests that the leaders of these villages were involved in organizing the construction and maintenance of canals.

The Pueblo Grande platform mound was initially constructed as two separate north and south mounds around 800 years ago. On the south mound, two pithouses were constructed, each facing away from the other (see platform mound map on page 30). These pithouses may represent two different kin or other social groups. Over time, the north and south mounds grew together and eventually became one large mound. Periodically, it was expanded and modified until it reached its present configuration and size.

The platform mound at Pueblo Grande is one of the largest the Hohokam ever constructed. Like the northwest ballcourt, it is oriented in a general north-south direction and measures 160 by 294 feet (49 by 90 m), approximately the size of a modern-day football field. At its greatest height, the mound stood 25 to 30 feet (7.5 to 9 m). Surrounding the entire mound, a 6 to 7 feet high (2 m) and 3 feet wide (1 m) compound wall restricted access to all but a selected group of individuals.

When Adolph Bandelier of the Archaeological Institute of America visited the site of Pueblo Grande in 1883 (see page 31), he painted a watercolor map of the platform mound. On top of the mound he depicted several smaller mounds. Archaeological evidence confirms that these were once rooms or structures and that numerous rooms had been periodically built and destroyed on top of the mound. In the 1930s, the south half of the platform mound was systematically excavated by Civilian Conservation Corps crews under the direction of Julian Hayden, revealing a maze of rooms, courtyards and other interesting architectural features. On the unexcavated north half of the platform mound one still can see the same mounds Bandelier painted over a century ago.

During excavation on top of the platform mound, an astronomically-oriented room was uncovered on the southeast corner of the mound. This room contains an unusual corner doorway located in the northeast corner of the room. In addition to this corner doorway, a center doorway was built in the south wall of the room.

In this room a very interesting event occurs on the summer solstice. At sunrise on the summer solstice a beam of light passes through the corner doorway and out the south doorway. This marks the longest day of the year. In contrast, the platform mound at Mesa Grande, a Hohokam village south of the Salt River, contains a room in which the winter solstice is marked. The Hohokam possibly used

Solstice Room on the Pueblo Grande platform mound. The Hohokam appear to have watched the sky and tracked the movement of the sun, moon and stars as these celestial bodies moved across the heavens. To mark certain astronomical events, the Hohokam built a special room on top of the Pueblo Grande platform mound. This room had two doorways, one in the northeast corner and another in the center of the south wall. At sunrise on the summer solstice, sunlight passes through the corner doorway and out the side doorway, marking the longest day of the year. The Hohokam may have used this knowledge for the scheduling of religious ceremonies. Similar features marking astronomical events have been found at several other Hohokam sites.

View of the Pueblo Grande platform mound with modern museum buildings in background, looking northwest. One of only a few platform mounds surviving into the modern era, the massive mound at Pueblo Grande is about the size of a football field and at one time stood over 20 feet (6 m) tall. Today, a fully accessible trail with interpretive signs takes visitors to the top of the platform mound and to the northwest ballcourt where they learn about the ancient Hohokam people. The museum buildings in the background were built in 1974 and 1995. Today's archaeological park is over 100 acres (40.5 ha) in size and preserves the center of the 1,500 year old village. To the south of the massive platform mound is an undeveloped portion of the park called the Park of Four Waters. This area contains the preserved remains of two large Hohokam irrigation canals.

A view of the northwest compound at the Pueblo Grande platform mound, looking southeast. The northwest compound is a maze of rooms, courtyards, storage rooms, outdoor cooking areas, and other features adjacent to the platform mound. These rooms and courtyards were probably used to support activities that were carried out on top of the platform mound. The only known access point through the compound wall was in this area. The remains of a two-story room can be seen on the left.

these seasonal markers to determine the planting time of crops, harvesting of wild plants, and to establish ceremonial cycles for important rituals.

Inside the northwest quarter of the Pueblo Grande platform mound compound, a group of adobe structures were constructed adjacent to the base of the mound. One of these rooms was two stories high. This maze of rooms has few access points, (i.e., doorways), suggesting that movement of people was limited among the rooms and unroofed courtyards. Artifacts recovered from this area include various minerals and pigments possibly used in ceremonial rituals, as well as tools used to make baskets and weave textiles.

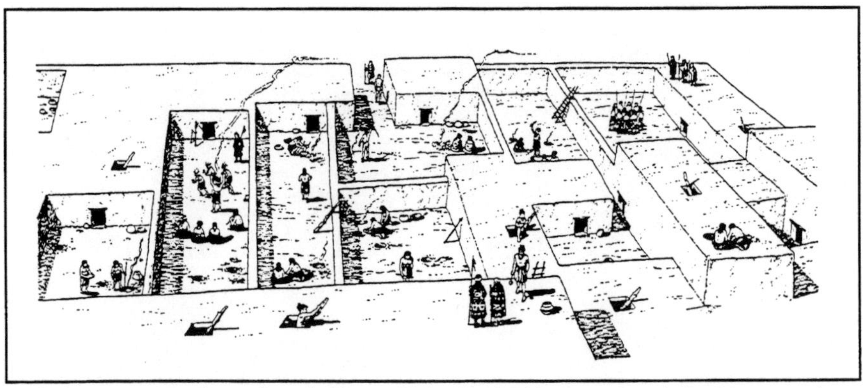

Artist's recontruction of activities within the NW compound, looking west.

The importance and prominence of the

Pueblo Grande platform mound over other Hohokam platform mounds can be seen in a number of ways. It was one of two of the largest platform mounds in the Salt River Valley, (Mesa Grande being the other), and the largest platform mound along Canal System Two. Because of its location at the headwaters of the canal system, it controlled the water in the majority of the canals along the north side of the Salt River. From its top, priests or elite leaders could observe as many as ten main irrigation canals in operation.

In the platform mound's construction, the Hohokam at Pueblo Grande used tons of locally quarried caliche, sandstone, and granite rocks to support the mound instead of relying exclusively on coursed-adobe, from which most other platform mounds were constructed.

Big Houses

A Hohokam big house was a large, multi-storied adobe tower-like structure built during the Classic period (A.D. 1150-1450). In 1887, Frank Hamilton Cushing, famous for his ethnographic work at Zuni Pueblo in New Mexico and leader of the Hemenway Southwestern Archaeological Expedition, reported the existence of a big house at Pueblo Grande. By the turn of the century, however, this structure had been destroyed. Although Cushing's notes are limited they suggest that the Pueblo Grande big house was somewhat similar to the four-story structure at Casa Grande Ruins National Monument in Coolidge.

A Hohokam big house was physically similar to a platform mound, and may have served similar purposes. Though they are

The big house at Casa Grande Ruins National Monument near Coolidge, AZ, ca. 1880s. (Photo courtesy of Casa Grande Ruins National Monument.) Big houses were constructed by the Hohokam during the Classic period in the early 1300s. Although somewhat similar in construction to platform mounds, big houses appear to be quite rare in Hohokam sites and may have had astronomical features. A big house at Pueblo Grande was excavated by Frank Hamilton Cushing in 1887, the first scientific excavation in the American Southwest.

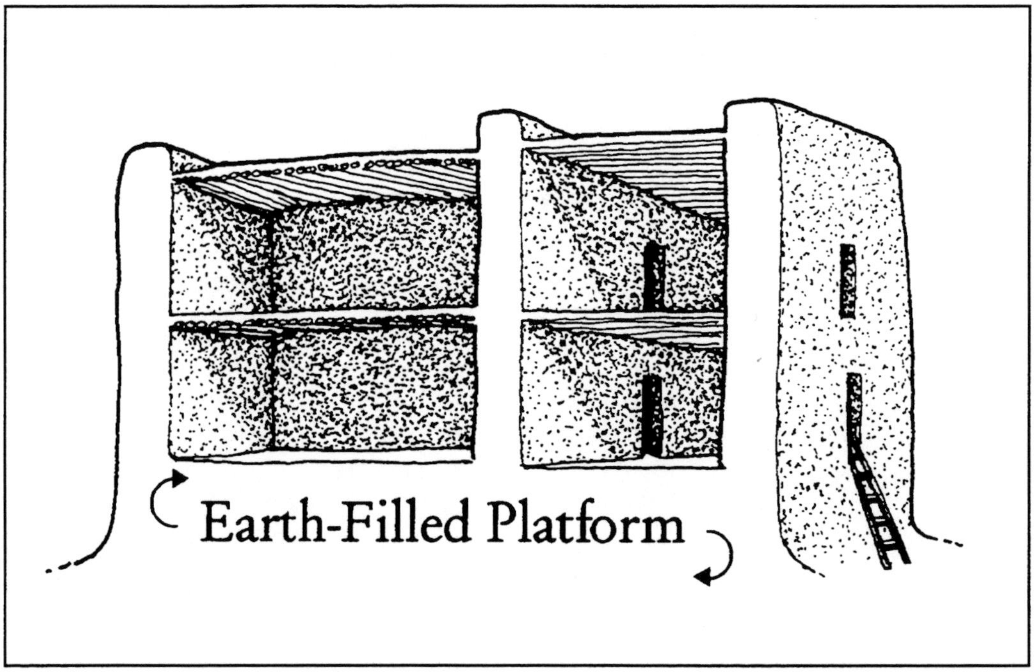

Artist's reconstruction of a Hohokam big house showing a cut-away view. Big houses were multi-story tower-like structures with multiple rooms. The bottom level of the Casa Grande big house was filled with dirt creating an earthen platform on which the subsequest floors were constructed. The walls of this big house were constructed of coursed-adobe and plastered with caliche. Wooden timbers were imported from miles away and used in the building of floors and roofs (much in the same manner as with house construction). Ladders are presumed to have been used to access the different levels.

multi-storied structures, the bottom story is a raised platform mound upon which the rest of the structure is built. Also, like platform mounds, these structures were enclosed by a compound wall.

Cushing reported the big house at Pueblo Grande as being at least three and possibly four stories high. The bottom floor had been filled (a platform mound) and subsequent stories were built on top, each floored and roofed. Each floor contained multiple rooms and the walls were covered with a fine white plaster. Inside one of the rooms the walls contained etchings of a human, animals, and lightning

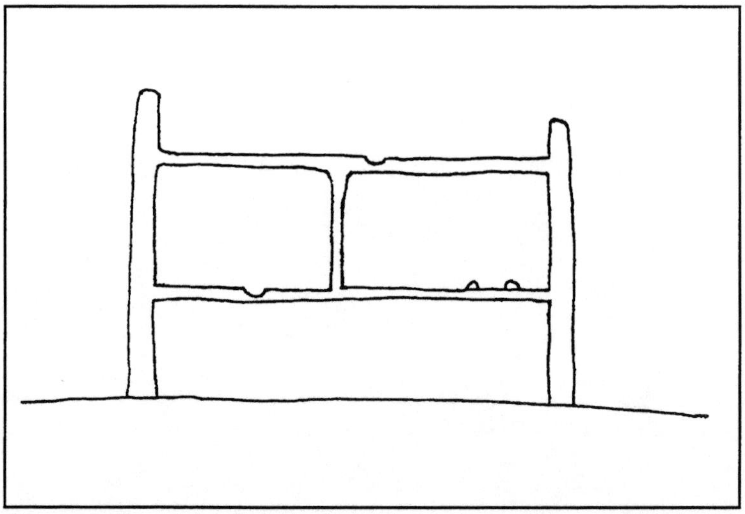

Cross-section of the Pueblo Grande big house (from 1887 sketch by Frank Hamilton Cushing). Architecturally similar to the big house at Casa Grande Ruins National Monument near Coolidge, the Pueblo Grande big house may have served, at least in part, as an astronomical observatory. Because of its poor preservation at the time of Cushing's investigation, however, no evidence of such function has been documented.

Etched wall designs inside the Pueblo Grande big house (recorded in 1887 by Frank Hamilton Cushing).

that possibly had ceremonial significance. According to Cushing, the Pueblo Grande big house appeared to have gone through at least two remodeling phases during its use. Cushing noted that it had been abandoned by the Hohokam and left to decay for a considerable amount of time before it was again rebuilt, and then finally abandoned sometime in the 1400s.

Due to its poor preservation at the time of Cushing's excavation, and its subsequent destruction, it is not known if the big house at Pueblo Grande was used as an astronomical observatory, as was the Casa Grande big house. The big house at Casa Grande Ruins contains several circular holes in its walls which are aligned with certain astronomical events such as solstices and equinoxes. Two holes in a third story room are aligned with the equinox and serve to mark the events with a ray of light. A hole in the west wall of a lower story room provides a view of the summer solstice sunset.

The existence of a big house at Pueblo Grande strengthens the argument that the site was very important to the Hohokam of the Salt River Valley. The Hohokam may have built several big houses, but only the Casa Grande big house has survived to the present day. This structure is approximately 60 by 40 feet (18 by 12 m) wide and 35 feet (10.5 m) tall.

Besides the big houses at Pueblo Grande and Casa Grande only two other possible big houses have been reported, both in the Salt River Valley, at the sites of La Ciudad and Las Colinas. Both are platform mound villages located within the same canal system as Pueblo Grande.

The location of the Pueblo Grande big house, about one-half mile (800 m) north of the platform mound, may indicate the existence of two distinct or separate political entities at Pueblo Grande during the latter part of the Classic period, or separate functions for these architectural structures.

Conclusions

The presence of numerous monumental or public structures in many Hohokam villages indicates that these desert farmers were so successful in their subsistence pursuits that they had time to build architecture beyond what was necessary for everyday living. Furthermore, this architecture probably symbolized important elements of their religion, although archaeologists are still unclear about the activities associated with these structures and which members of their society participated in the activities.

Hohokam artisans. Hohokam artisans manufactured a wide variety of utilitarian as well as ceremonial crafts. Most craft activities were probably done outside under open air ramadas. In this illustration, several manufacturing activities can be seen. The individual to the right is making pottery and is wearing shell bracelets on her upper arm, while the individual in the center is producing a stone axe.

Craft Production and Artistry

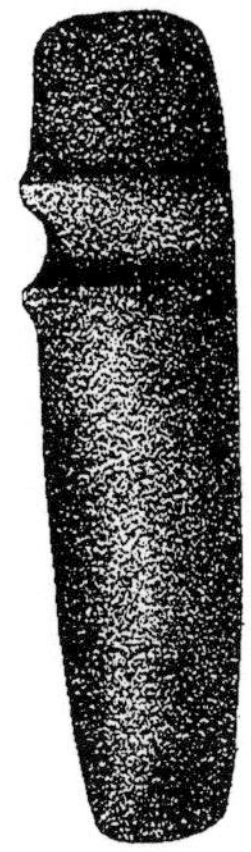

Three-quarter grooved stone axe from Pueblo Grande.

The Hohokam were talented artisans and craftspeople, some of the finest in the Southwest. They were expert workers of stone, bone, shell, clay, and possibly wood. From material obtained locally and by trade, the Hohokam made an assortment of tools, utensils and ornaments. Their beautiful designs were artistic as well as functional.

Most tools and ornaments made by the Hohokam were manufactured for personal or family use. Some individuals or possibly communities, however, may have acted as part-time specialists. A few highly regarded sources of raw material may have been controlled by certain groups, or an individual may have been particularly adept in the manufacture of certain items.

Lithic Tools

The Hohokam did not fashion metal tools so they relied predominantly on *lithic* (stone) material. Lithic tools are divided into two categories based primarily on manufacturing techniques. These two tool categories are chipped and ground stone.

Hohokam chipped stone tools were made from lithic material such as chert, chalcedony, quartzite, basalt, obsidian and jasper. This material was used for the production of knives, scrapers, choppers, hoes, drills, and projectile points. In

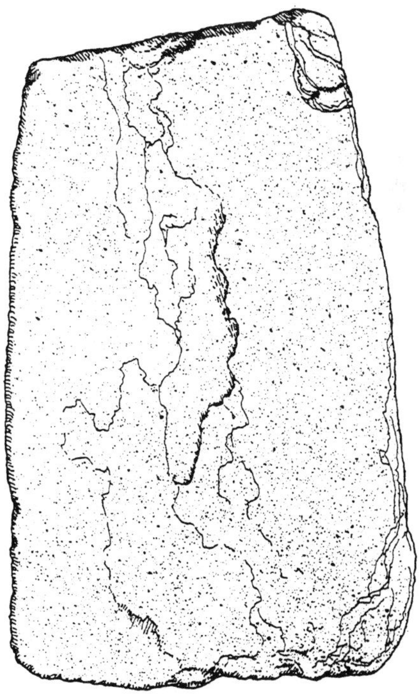

Tabular schist knife from Pueblo Grande. These tools are often called agave knives because of their use in the procurement of desert plants, in particular the harvesting of agave hearts and leaves.

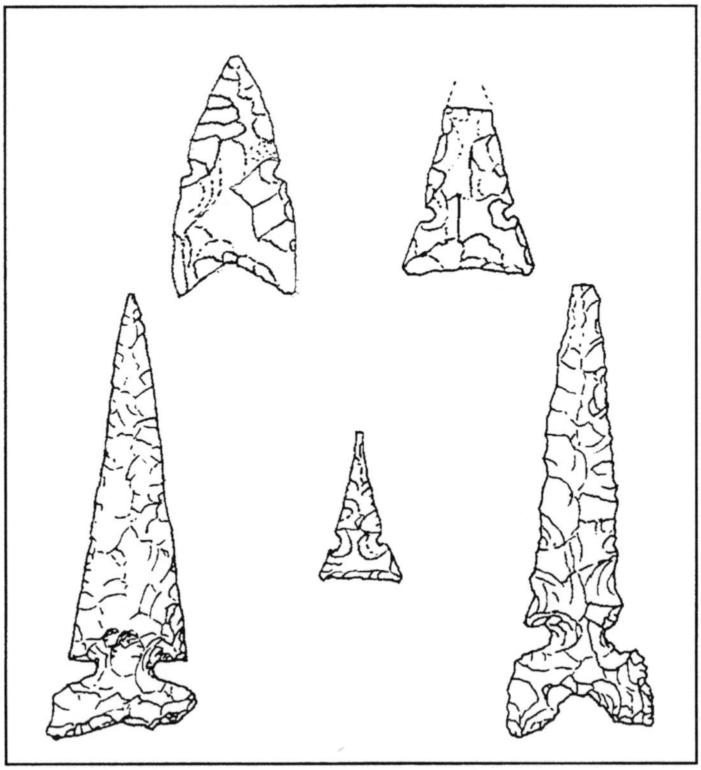

Projectile points from Pueblo Grande. Hohokam projectile points were generally triangular in shape with hafting notches in their sides. Projectile points found at Pueblo Grande were predominately made from obsidian originating from 10 different sources up to 140 miles (225 km) away. (Actual size.)

addition, large hand-held tabular schist knives were used to harvest wild plants.

At Pueblo Grande the Hohokam preferred to use obsidian to make their *projectile points* (arrowheads), although chert was also popular. Nearly half of all projectile points recovered from the site were manufactured from obsidian, a volcanic glass. Obsidian recovered from Pueblo Grande originated from 10 different sources, ranging in distance from 50 to 140 miles (80 to 225 km) from the site. The most favored obsidian came from the Sauceda Mountains, located 69 miles (111 km) to the southwest of Pueblo Grande, near Gila Bend.

Early Hohokam projectile points are some of the most elaborate and beautiful points ever manufactured in the Southwest. They are relatively large in size with barbs or deeply serrated edges and would have made formidable-looking weapons. These projectile points were clearly made by specialists. Later Hohokam projectile points were much smaller in size and usually triangular shaped.

Hohokam ground stone tools were manufactured from porous lava material such as vesicular basalt, used to fashion *manos* and *metates* (food grinding tools). Highly polished stone axes were manufactured from diorite and other river cobbles. These axes were hafted with wooden handles and were effective in chopping down hard desert trees, such as mesquite, for construction material. Other ground stone lithic tools included mortars and pestles, polishing stones, hammers, plummets, hand files, and arrow shaft smoothers.

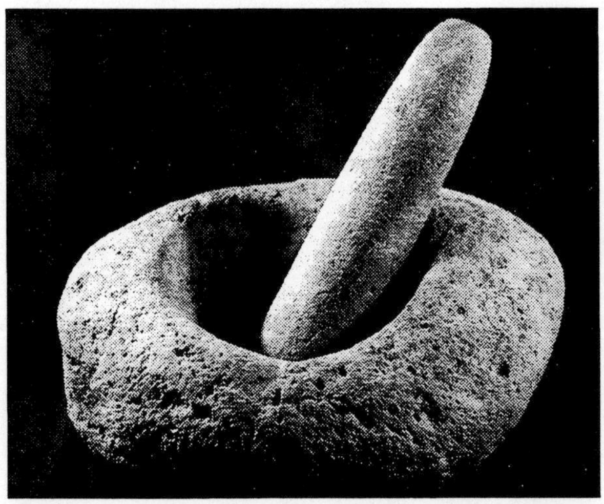

Mortar & Pestle. Mortars (base stone) & pestles were used by the Hohokam and other peoples of the Southwest to grind seeds and seed pods into flour. These tools were particularly useful in the processing of mesquite bean pods. Mesquite bean pods are thought to have been a staple for the Hohokam people.

Grinding corn using a mano and metate. A mano (being gripped in the picture), is a flat hand-held stone used in conjunction with a metate (base stone). Hohokam metates were generally trough shaped as the one shown here. Manos and metates were used for grinding dried corn and seeds into meal or flour. These tools were widely employed by cultures of the Southwest and date back thousands of years.

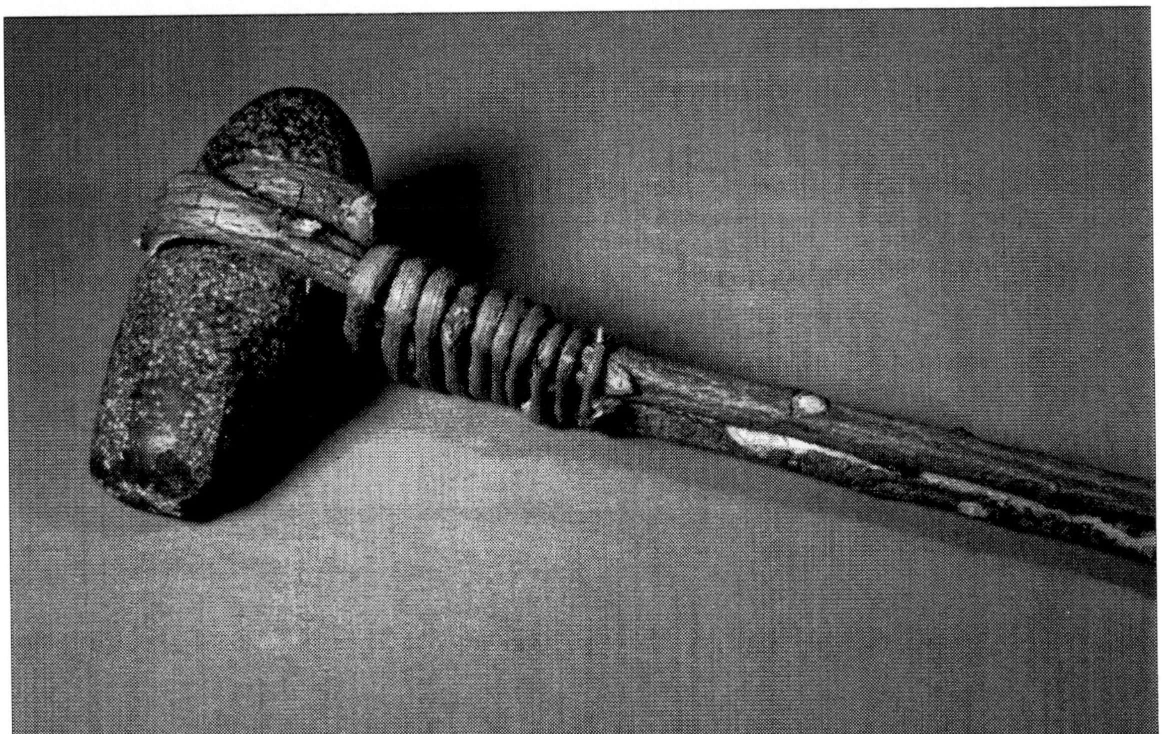

Three-quarter grooved axe with replicated haft. This style of manufacture is typical of Hohokam axes which would have been hafted with a wooden handle wrapped around the groove. Hohokam axes were made from stone such as diorite or other river cobbles. These axes were very effective in chopping down hard desert trees to be used as firewood or building material.

Shell Jewelry and Ornamentation

The Hohokam were master craftspeople of marine shell jewelry. At Pueblo Grande, shell ornaments were worn by both males and females, adults and children. Shell jewelry was often buried with the deceased, probably possessions once owned by the individual.

More than 30 species of shell were used by the Hohokam, primarily imported from the Gulf of California, but also from the Pacific Coast. Commonly used types of shell were the *Glycymeris gigantea*, *Laevicardium*, *Olivella dama*, *Pecten*, and *Conus*. Thousands of shell ornaments were recovered at Pueblo Grande during excavations. Although some shell jewelry was made at the site, it does not appear that Pueblo Grande was a major manufacturing or redistribution center.

The Hohokam used several techniques or processes in the fashioning of shell jewelry. One technique consisted of etching, which probably was done by painting the desired design on the shell using a resin or pitch, then immersing the shell in juice from the saguaro fruit or other plants.

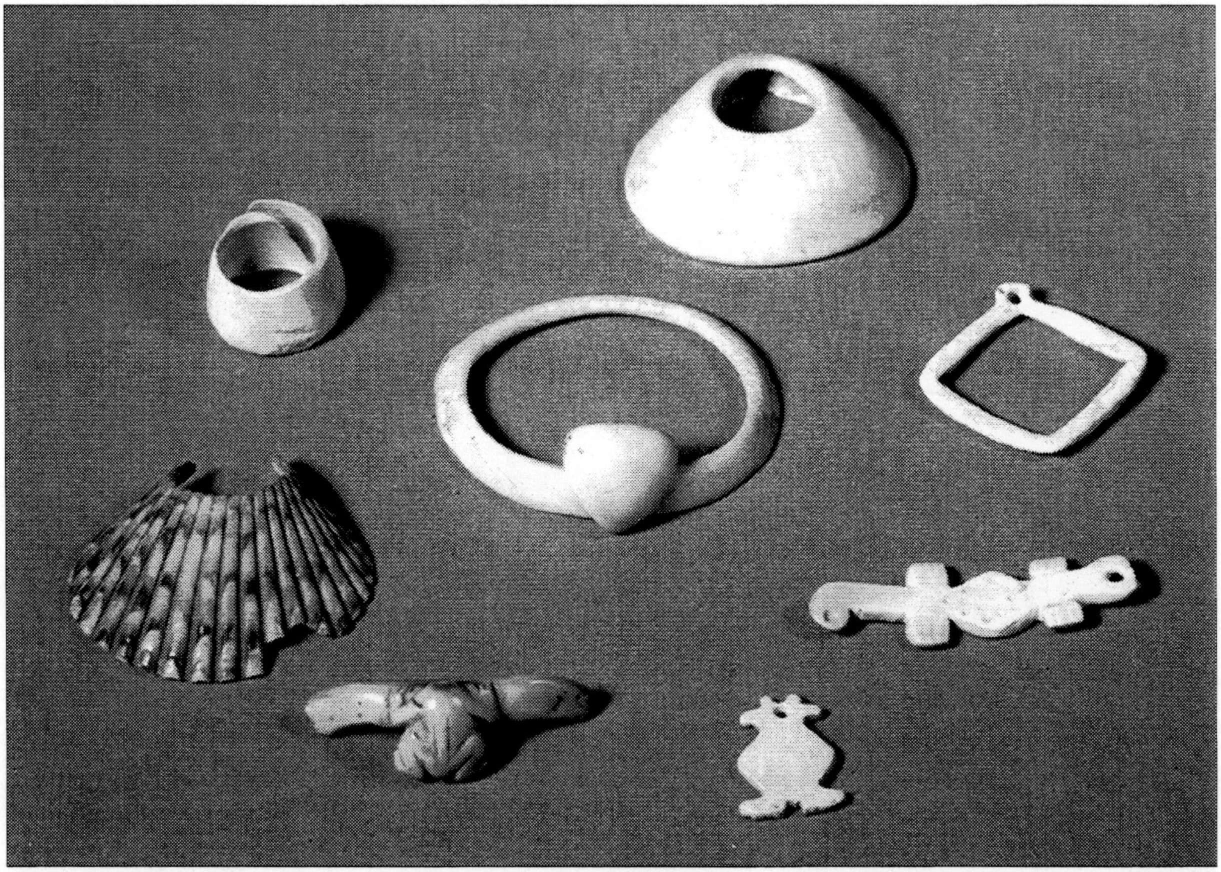

Hohokam shell jewelry. Shell was obtained through trade from hundreds of miles away, mostly from the Gulf of California and the Pacific Ocean. Lizards and frogs, as well as water birds, seem to have held special significance for Hohokam jewelers, perhaps because of their relation to water. Thousands of pieces of shell jewelry were excavated from Pueblo Grande, where they were worn by men, women and children. The Hohokam also traded shell jewelry to other people of the Southwest to obtain many of the raw materials not readily available to them, as well as to form relationships with their neighbors. Although jewelry was made at Pueblo Grande, the village does not appear to be a major manufacturing or redistribution center.

The acidic juice ate away the unprotected portions of the shell, leaving the protected areas in raised relief. The Hohokam were also proficient in the cutting and grinding of shell to shape it into rings, bracelets, pendants, *tinklers* (small bell-like objects fashioned from marine shell), and animal figures (especially frogs, lizards and birds). Small shells were drilled and strung together as beads to make necklaces and bracelets.

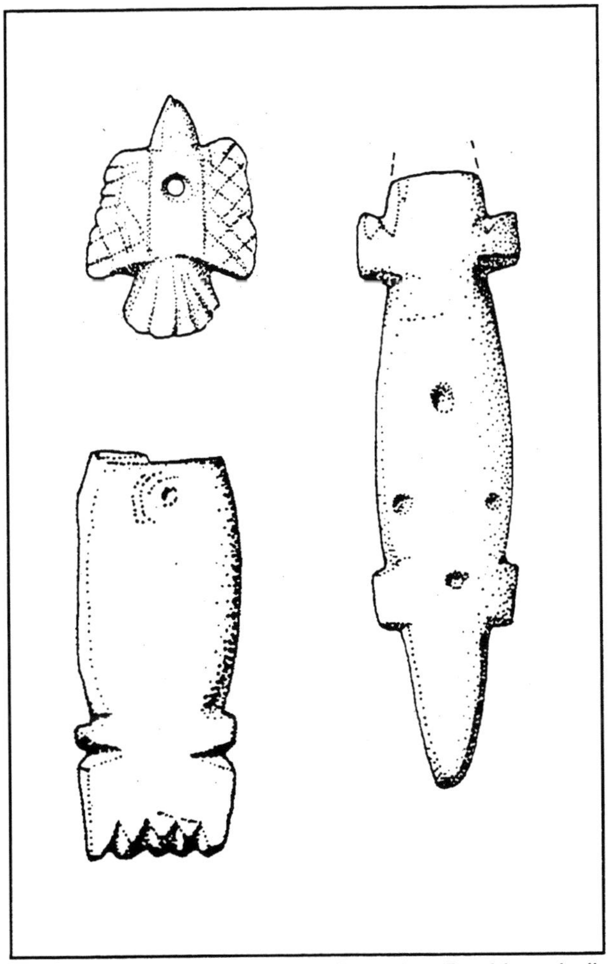

Stone jewelry from Pueblo Grande. Besides shell, the Hohokam also made much of their jewelry out of local and imported stone. The pendant on the lower left is made from schist while the bird pendant on the upper left is made of serpentine. The lizard on the right is made of argillite. The depressions carved into the pendant's back may have once contained turquoise. (Approximately 2/3 actual size.)

Stone Jewelry

The Hohokam also made jewelry out of stone. From argillite they manufactured beads, rings, overlay, lip and nose plugs, and effigies. Argillite artifacts at Pueblo Grande came from three different sources: the Upper Verde Valley, the Upper Tonto Basin, and possibly the Tucson Basin.

Turquoise was another type of stone highly prized by the Hohokam, as it is today by most Southwestern cultures. Turquoise at Pueblo Grande had been crafted into beads, pendants and *tesserae* (mosaic pieces). This turquoise came from at least five different sources in eastern Arizona and New Mexico. In addition, turquoise from southern California has been recovered from the site of Snaketown, south of Pueblo Grande along the Gila River.

Textiles

The Hohokam made a variety of clothing from woven cotton, such as blankets, breech cloths, skirts or kilts, hats or turbans, and shirts. Actual specimens of textiles obtained from archaeological sites indicate that the Hohokam possessed knowledge of a complex weaving technique. Other evidence comes from the recovery of numerous ceramic and stone spindle whorls.

Wooden spindles were used by the Hohokam to spin cotton into yarn. The spindle was probably about a foot (35 cm) long and would have been rolled on the leg or thigh. A small perforated ceramic or stone whorl (disk) attached to the spindle acted as a counter weight.

For everyday clothing, Hohokam men probably wore cotton loincloths while women wore simple skirts. Blankets were

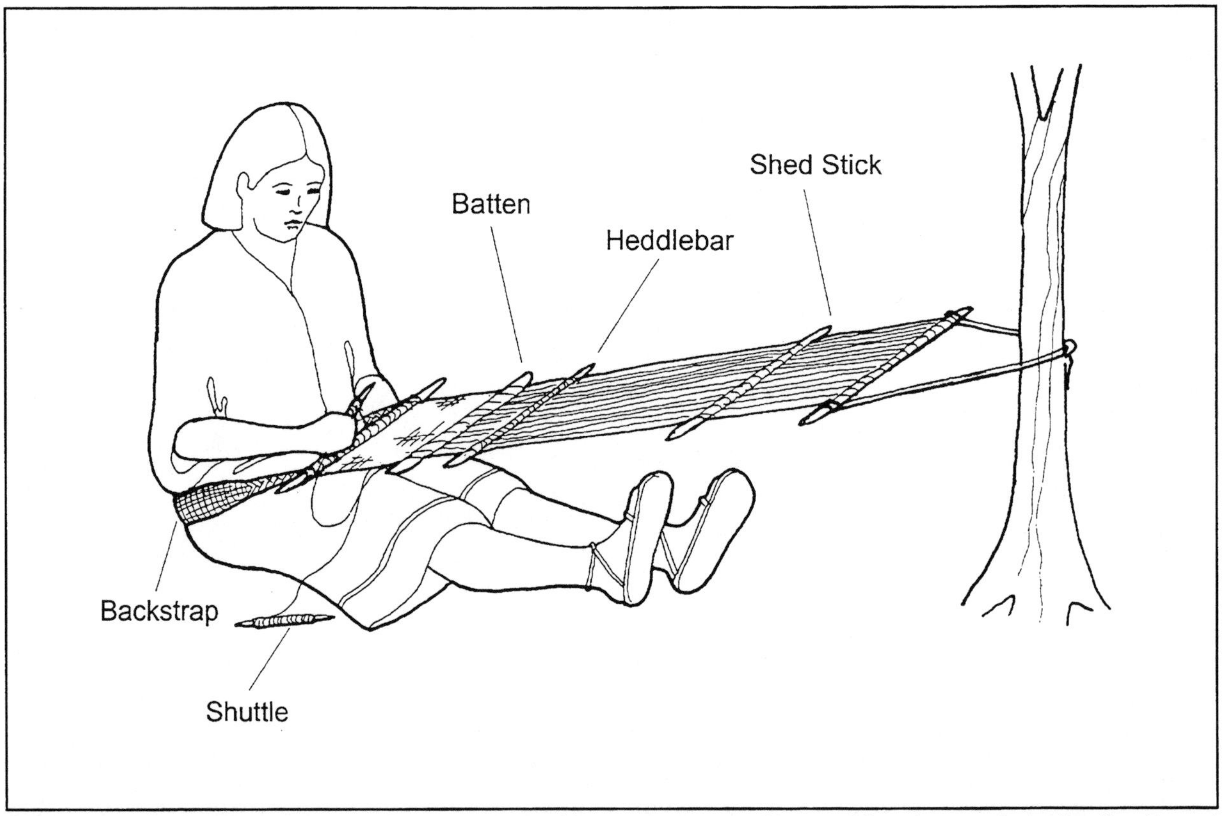

Illustration of a backstrap loom. Several types of looms were used by prehistoric people in the Southwest including the Hohokam. Although no looms have been found in Hohokam sites, they are believed to have used versions of the backstrap (shown above) and horizontal style.

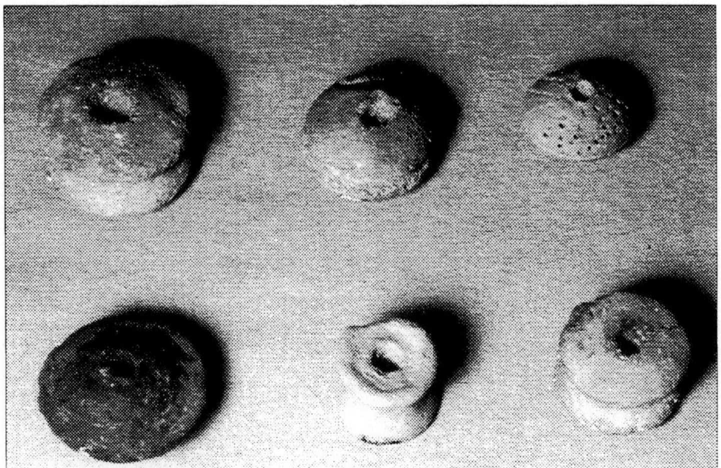

Modeled ceramic spindle whorls from Pueblo Grande. Spindle whorls were often made from shaped clay, pottery sherds, and stone. These artifacts are very common in Hohokam sites. The site of Pueblo Grande has yielded hundreds of spindle whorls, which suggests that the people of the village were involved in textile manufacture for possible trade with other Hohokam villages or neighboring peoples.

probably employed to keep out the cold during the cooler months.

Besides cotton, the Hohokam used tough agave fibers to weave belts and ropes. Other plants provided the Hohokam with material for sandals and plaited mats. At Pueblo Grande, remains of reed mats, possibly used as sleeping mats, were found in several rooms on top of the platform mound.

Although Hohokam textiles did not preserve well in the archaeological record, remnants of fabric found in caves, and woven impressions left in clay have been recovered from Pueblo Grande and other Hohokam sites.

Wooden Tools

Another type of artifact that does not preserve well in the archaeological record is wood. The Hohokam are believed to have used wood for a number of purposes in addition to building material, such as digging sticks, paddles, handles and other objects. Wooden artifacts have been found at Pueblo Grande, including several wooden staffs or wands associated with male and female burials. Three of these staffs were painted blue, and may have represented some kind of badge or status marker.

Bone Tools and Musical Instruments

The Hohokam also used animal bone to make some of their tools. For example, bones were made into awls and used for basketry and leather working. At Pueblo Grande, most awls were made from deer leg bones, but also included rabbit, dog, and rare raccoon bones.

In addition to utilitarian tools, the Hohokam fashioned musical instruments out of animal bones. Whistles were made from bird bones, and other instruments included flutes, rasps, and tortoise shell rattles. These items were likely an important aspect of Hohokam daily life and ceremonial activities.

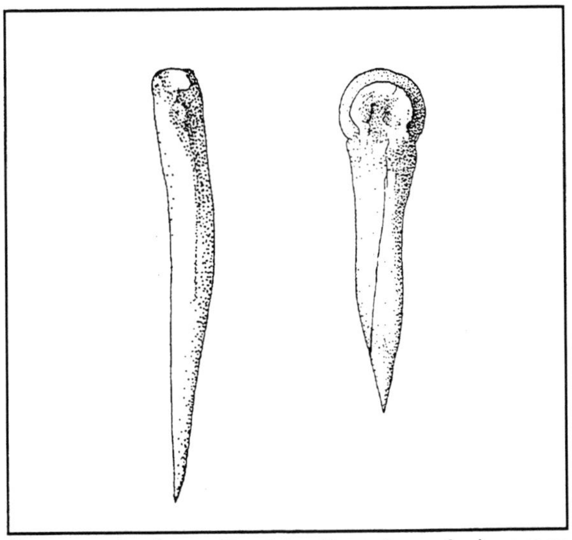

Bone awls from Pueblo Grande. Awls were generally made from deer leg bones and were used as tools in the manufacturing of baskets and leather goods.

Basketry

Before and after the introduction of pottery, the Hohokam used locally available plant fibers to weave baskets. Hohokam baskets were typically coiled and made from willow fibers and arrowweed. Pieces of charred basketry have been found at sites such as Snaketown and Pueblo Grande. Baskets probably were used for a variety of purposes in which heavier or more fragile pottery was unsuitable. Some archaeologists have theorized that lining baskets with clay led to the development of pottery.

Ceramics

The most common type of artifact indicative of the Hohokam culture is their pottery. Pottery was manufactured from local clays mixed with sand, crushed rock and other temper to prevent the clay from cracking and breaking during the firing process. Hohokam pottery was generally of three types: plain, red, and the widely known red-on-buff decorated wares. Pottery also occurred in a wide variety of shapes and sizes, including jars, bowls, pitchers, scoops, and plates. Sizes of vessels ranged from miniatures to large jars, called *ollas*, some of which could store enormous amounts of water (25 gallons

Snaketown Red-on-buff pottery. The Hohokam manufactured many types of pottery, but the most widely recognized is the red-on-buff. This type of pottery was first manufactured starting around A.D. 650. Hohokam artists created a myriad of animal and geometric designs using painted red lines over a buff colored *slip*.

[94.5 l] or more). The Hohokam decorated their pottery with a multitude of designs. Many of these designs were geometric, but the Hohokam also used representations of animals as well as humans.

Hohokam pottery vessel technology most likely developed when increased dependence on agriculture necessitated the production of watertight, pest-resistant containers for storage, and thermal-shock resistant pots for cooking food directly over a fire. In building a temporal chronology for the Hohokam cultural sequence, archae-

Plainware pitcher from Pueblo Grande.

ologists have often relied on ever changing pottery styles. Changes in Hohokam pottery styles were mostly in decoration (i.e., color of paint and slip), although some changes occurred in the manufacturing process as well. Recently, archaeologists have been studying the types and sources of temper in Hohokam plainware pottery to better understand exchange relationships among Hohokam villages.

Current data indicate that Hohokam plainware pottery was produced by A.D. 1, and continued to be crafted throughout the Hohokam occupation of the Sonoran Desert. Redware pottery was first made sometime after A.D. 400, and the first painted pottery, red-on-gray (predecessor to red-on-buff), was manufactured ca. A.D.

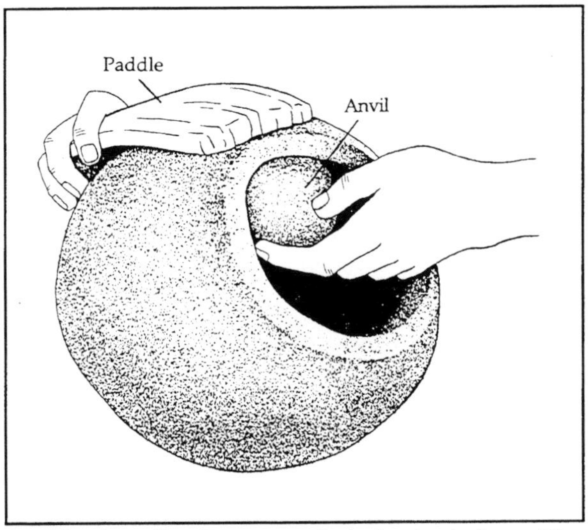

Paddle & anvil pottery manufacturing technique. The Hohokam used the paddle & anvil technique when making their pottery. The vessel is shaped with rolled coils of clay and then is smoothed with the use of a paddle.

Sacaton Red-on-buff vessel with a Gila Shoulder.

Hohokam pottery designs from Pueblo Grande. These designs, taken from pottery sherds, show the wide variety of Hohokam ceramic motifs.

650. By the Colonial period (ca. A.D. 750), the Hohokam were creating beautifully painted vessels. Pottery was decorated with red paint over a buff colored *slip*, or thin outer clay covering. Colonial period red-on-buff pottery was expertly crafted and artistically decorated with a myriad of geometric designs as well as images of humans, mammals, reptiles, fish, birds, and even flowers.

Late Pioneer and Colonial period potters also incised a series of parallel encircling grooves on the outside of the vessel on top of the painted designs. Colonial period pottery recovered from Hohokam sites, including Pueblo Grande, is painted with decorations which depict human figures performing various activities. Humans hold hands and dance, hunched figures with *tumplines* around their foreheads carry burden-baskets on their backs, plumed hunched-back flute players (which may represent the mythical Kokopelli) play music, and humans hold bows and arrows that face animals such as big horned sheep and deer. All were

Bird effigy vessel from Pueblo Grande. The Hohokam created a wide variety of ceramic vessels with animal characteristics.

Redware ceramic bowl with legs and feet from Pueblo Grande. These unusual Hohokam bowls were supported by three and sometimes four legs. This bowl's legs even have feet to help support it.

masterfully painted with red lines on buff colored background ceramic pots. Other pottery depicts creatures from the natural environment of the Hohokam, and all are set in fluid motion by the artist. Schools of fish swim, dogs or coyotes stand alert with heads up and tails raised, coiled rattlesnakes are ready to strike, turtles crawl, lizards creep, flocks of birds fly, and water birds hold fish in their beaks.

During the Sedentary period (ca. A.D. 900-1150), pottery vessels may have been mass produced, suggesting an increasing importance

Redware ceramic vessels. Redware vessels were made early and late in the Hohokam culture, but were especially common during the Classic period. These highly polished vessels were often used as funerary objects and may have contained food, water or other materials.

Gila Polychrome pot from Pueblo Grande collection. Polychrome pottery became prevalent during the Classic period when the Hohokam imported or imitated Salado pottery from the east in the Tonto Basin. Vessels came in a variety of shapes and colors (i.e.,black, white and red).

on quantity of production. By the 12th century, the quality of Hohokam red-on-buff pottery had deteriorated considerably, showing the decline of a long ceramic tradition. At this same time, Hohokam potters developed their skills in making polished redware ceramics in a variety of pleasing shapes and sizes. Redware vessels were

Polychrome ceramic sherd from Pueblo Grande.

Painted ceramic animal figurines from Pueblo Grande. These possible dog effigies were part of a cache of seven figurines excavated from a Sedentary period pithouse floor. Similar caches have been excavated from other Hohokam village sites, most notably, the site of Snaketown along the Gila River.

accompaniments, and may have been traded between villages. In addition, the Classic period Hohokam imported and imitated relatively large quantities of polychrome vessels (i.e., black, white and red colors) from the Salado cultures to the east in the Tonto Basin. These changes in pottery styles coincide with a general change in Hohokam culture during the Classic period.

Effigies and Figurines

Another form of artistic expression of the Hohokam was the manufacturing of clay and stone effigies and figurines. The first ceramic effigies in the Southwest were manufactured during the Desert Archaic period around 800 B.C. Hohokam effigies included humans as well as animals.

Anthropomorphic or human effigies, both male and female, were made in many forms and show important details about the Hohokam people. Figurines often have body decorations, including body paint, and show clothing styles, all of which may be renditions of actual Hohokam people. Some figurines are depicted with shell and stone jewelry. Others wear what appear to be arm and shoulder pads, possibly representing Hohokam ballplayers.

Zoomorphic effigies consist of animals found within the Hohokam environment such as dogs, mountain sheep, and deer. A cache of seven ceramic dog effigies was found on a Sedentary period (A.D. 900-1150) pithouse floor at Pueblo Grande. Human and animal effigies may have been merely an outlet for artistic expression, or more likely had religious or ceremonial significance.

Black-on-white trade vessel found at Pueblo Grande. This vessel, excavated from Pueblo Grande, was a trade item from northern Arizona.

A Flow of Goods and Ideas

Shell lizard pendant from Pueblo Grande.

As with other cultures, the Hohokam traded for items not readily available to them. The Hohokam trade network was vast and stretched from Mexico to Utah, from the Pacific Coast to New Mexico, and into the Great Plains. The Hohokam traded for a number of raw materials and finished craft products, and may have obtained these items by trading their cotton, surplus crops, and their shell jewelry. Hohokam ceramics have been found at many non-Hohokam sites in small quantities. Hohokam vessels themselves, however, do not appear to have been popular trade items, but may have contained food or other desirable materials.

Trading for Resources

The Hohokam traded shell obtained from two different locations: warm-water shells from the Gulf of California and cold-water shells from the Pacific Coast (e.g., abalone). Upon obtaining these shells, either through trade or by collecting the shells themselves during long distance trips, the Hohokam created beautiful ornaments which they used for their own personal adornment and also as trade items. Hohokam shell jewelry, especially carved bracelets and arm bands, has been found in numerous sites occupied by neighboring cultures.

Other items the Hohokam people traded for included turquoise, obsidian, argillite, steatite, and other minerals, as well as a multitude of foreign pottery types. At Pueblo Grande, more than 50 types of imported ceramic wares have been recovered during excavation. This pottery was imported throughout Pueblo Grande's occupation and came from at least five different cultural regions reaching out in all four cardinal directions: the Anasazi (ancestral Pueblo) and Hopi in northern Arizona, the Prescott area to the northwest, the Mogollon and Mimbres to the east, the lower Colorado River to the west, the Tucson Basin to the south, the Trincheras region of northern Sonora, Mexico, and Casas Grandes to the southeast in northern Chihuahua, Mexico. Interestingly, most of the imported pottery types were small bowls, possibly because bowls could be stacked and were therefore less likely to break during transport.

From Mexico the Hohokam also obtained macaws, obsidian, pyrite mirrors and copper bells. They did not have the

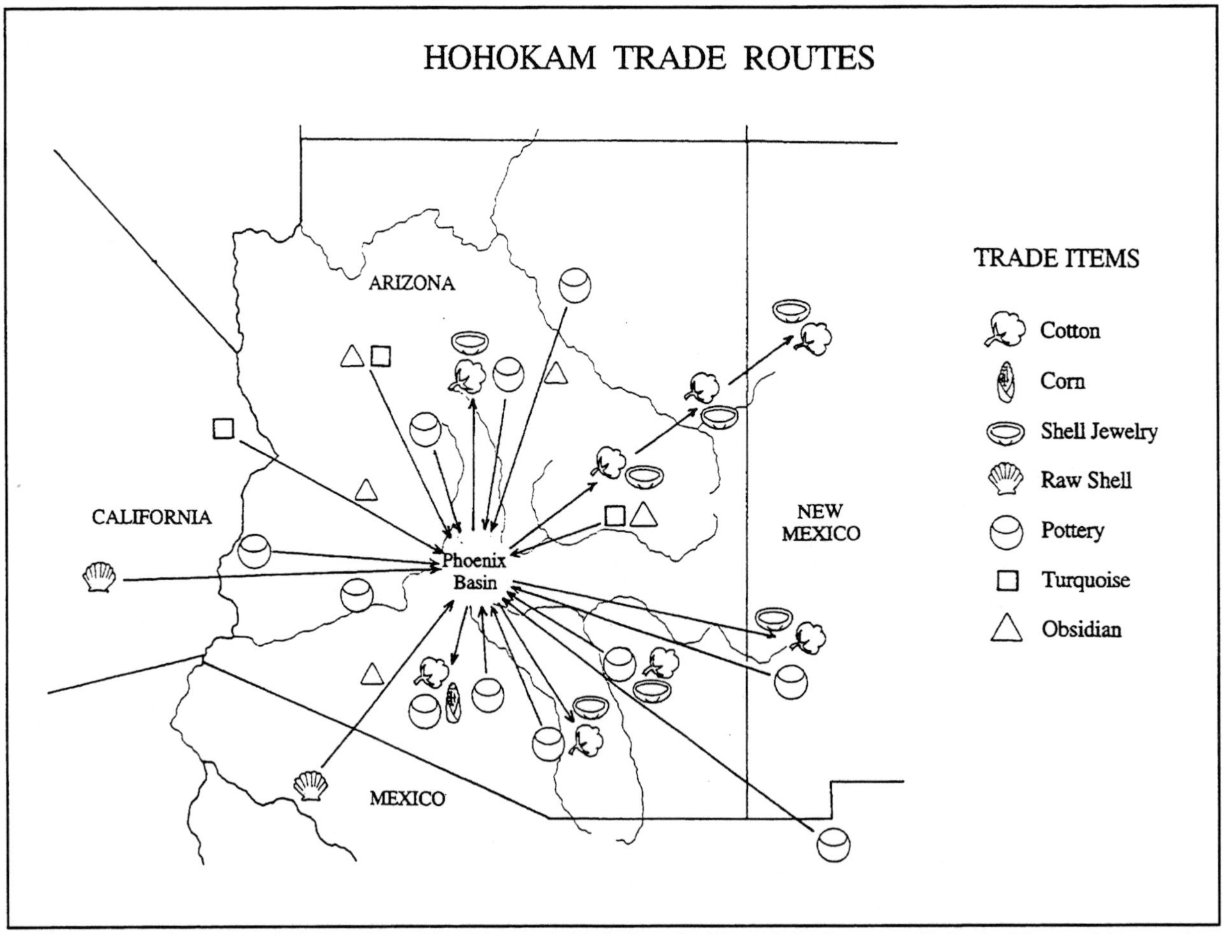

Hohokam trade network (after Doyel in *Chaco & Hohokam*). The Hohokam participated in a wide ranging trade network that stretched for thousands of miles from Mexico to northern Arizona and from California to New Mexico.

knowledge to make metal ornaments and tools, therefore, copper bells must have been very valuable trade items. At Snaketown, a Sedentary period (A.D. 900-1150) pithouse contained 28 copper bells, which may have been strung together into a necklace. At Pueblo Grande a single copper bell was discovered in a trash mound next to the platform mound by Erich Schmidt in 1924. More recently, eight copper bells were found together about a half mile north of the platform mound in 1999. Copper bells have also been found at several other Hohokam sites along the Salt River such as Las Acequias, Cashion, La Ciudad, and Los Hornos.

The Hohokam also traded for macaws and parrots. These birds, although relatively rare at Hohokam sites compared to other Southwestern cultures, were probably used for their colorful feathers in ceremonial and ritual activities. Their exotic nature would have been highly prized by the Hohokam.

The Flow of Ideas

Besides material goods, the Hohokam trading network also would have brought cultural influences. Hohokam ideology appears to have changed over time and

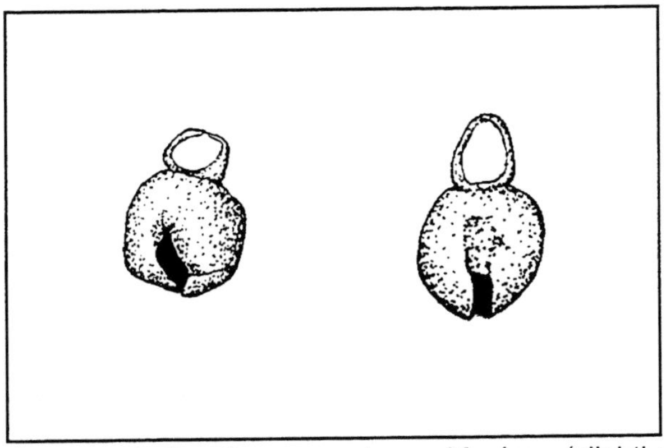

Copper bells from western Mexico (slightly enlarged). Copper bells were imported as trade items and may have served as status markers or indicators of elite prestige among the Hohokam. A number of copper bells have been discovered at Pueblo Grande. One of the bells was excavated from a large trash mound near the platform mound.

was probably influenced by contacts with other cultures through trade. Hohokam material culture indicates that they participated in religious and social status activities with groups outside their sphere of influence, including Mexico. In addition, the building of ballcourts with their ceremonial or ideological significance, the building of platform mounds, and changes in mortuary practices all may indicate cultural influences on the Hohokam world view through participation in a large interaction sphere.

Conclusions

The role of trade in Hohokam society has always been an important topic of discussion for archaeologists. The focus of this discussion has been on long-distance trade and the influence of Mesoamerican culture on Southwestern economy, but Hohokam trade occurred at many levels and involved both distant and local groups.

Although the variety of trade items from Mexico is interesting, they occur in relatively low frequencies over several centuries. It is possible that some of the exotic or finely made trade items served as status markers or indicators of elite prestige among the Hohokam. Some of these craft objects also may have been important in ritual transactions validating major life transitions such as marriage, completion of an apprenticeship, or receipt of power.

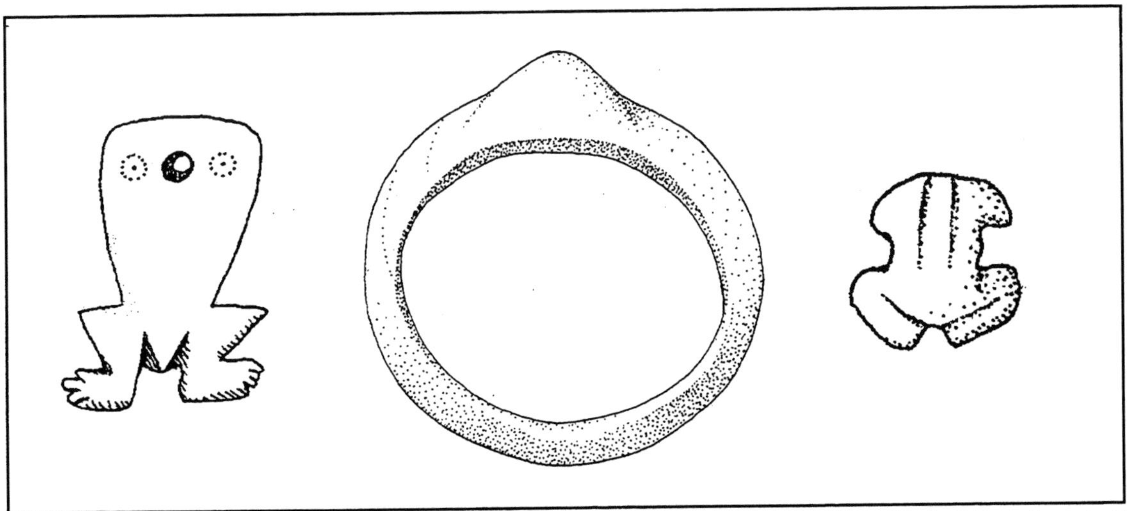

Shell jewelry from Pueblo Grande. The Hohokam created beautiful shell jewelry from marine shell obtained from the Pacific Ocean and the Sea of Cortez. Hohokam shell jewelry was a widely traded item and has been found at numerous sites of neighboring Southwest cultures.

Artist's conception of a Hohokam ceremonial scene. Individual on left is playing a reed flute. Dancers with gourd rattles are wearing macaw tail feather headdresses. Observers are wearing carved shell jewelry and woven cotton clothing.

Ideology and Worldview

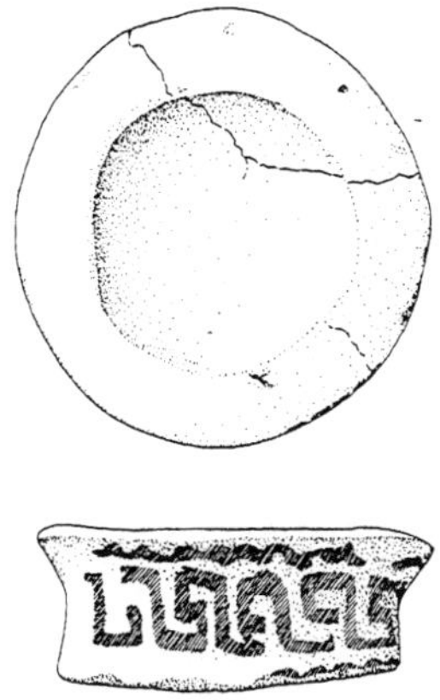

Ceramic censor from Pueblo Grande.

Southwestern American Indians are well known for their rich and complex religious systems. Archaeological data suggest that prehistoric cultures such as thc Hohokam probably also had an elaborate belief system, which was expressed in their public architecture (ballcourts, platform mounds, and big houses), material culture (palettes, censers, stone and shell jewelry, and pottery motifs), mortuary practices, and rock art. Hohokam ideological beliefs and practices set their culture apart in some respects from other prehistoric groups. Hohokam ideology probably was influenced by cultures from Mesoamerica, as well as by other neighboring Southwestern cultures.

Archaeologist David Wilcox has identified a Hohokam-Mexican exchange route that he calls the *Tepiman Connection*, a series of geographic areas joined by related Uto-Aztecan languages. This connection may have facilitated the flow of ideas as well as trade between the Hohokam and cultures from Mexico.

Wilcox has also proposed that some Hohokam priests may have made journeys to western Mexico to apprentice themselves with the well developed religious organizations there. Upon completion of their training, they may have brought back Mexican artifacts, as well as ideas, as symbols of their new status.

Artist's conception of two ceremonial rooms on top of the platform mound at Pueblo Grande. Many of the rooms on top of the mound are thought by archaeologists to have been used for ceremonies. The front room shown here had eight foot (2.5 m) high ceilings, very unusual for Hohokam architecture.

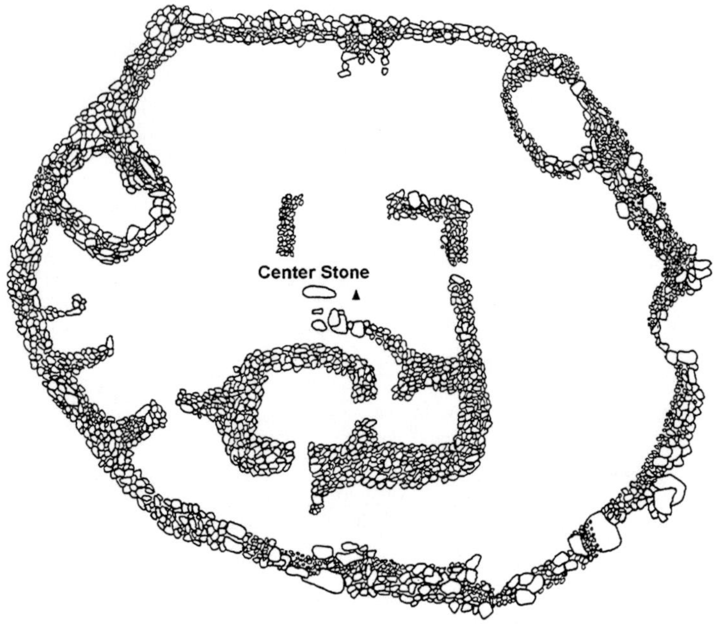

Shaw Butte hilltop site in northern Phoenix. Once thought by archaeologists to be a hilltop fortress, evidence now shows that the Shaw Butte site may be an astronomical observatory built by the Hohokam over 600 years ago. Several astronomical events related to the movement of the sun have been observed at the site. Alignments and light patterns mark numerous seasonal events such as summer and winter solstice sunrises and sunsets. These observations may have been used by Hohokam leaders for establishing ceremonial cycles for defining and sanctifying social behavior.

Mortuary Practices

Archaeologists argue that patterned distinctions documented in the study of a culture's *mortuary* or burial practices are a direct reflection on the living population. These patterns may represent differences in social status within the group as well as the culture's worldview or cosmology. Among other things, archaeologists look at how the body was prepared before burial, whether the body was cremated or inhumated, the position in which it was placed in the grave, artifacts or offerings placed with the body, and where it was located within the larger context of the site. The study of mortuary practices at Pueblo Grande, through the excavation of over 1,000 burials, has led to some interesting conclusions about the people who once lived there.

The Hohokam people at Pueblo Grande and elsewhere used two different mortuary practices to care for bodies upon their death. Like present day American society, the Hohokam both cremated and buried (inhumated) their dead. Also, like many societies around the world, past and present, the Hohokam often placed funerary items such as pottery, jewelry, and other possessions with their dead and most likely performed ceremonial death rituals and rites of passage.

The Hohokam used the cremation burial method predominantly during the Pioneer through Sedentary periods (A.D. 1-1150), but also to a lesser extent in the Classic period (A.D. 1150-1450). Because cremation is a destructive process, it is difficult to know with certainty the physical characteristics of the Hohokam people during most of their early cultural history. What has been learned comes from the few inhumations dating to the Sedentary period (ca. A.D. 900-1150).

Over 300 cremation burials have been recovered from Pueblo Grande excavations. After cremation in a large

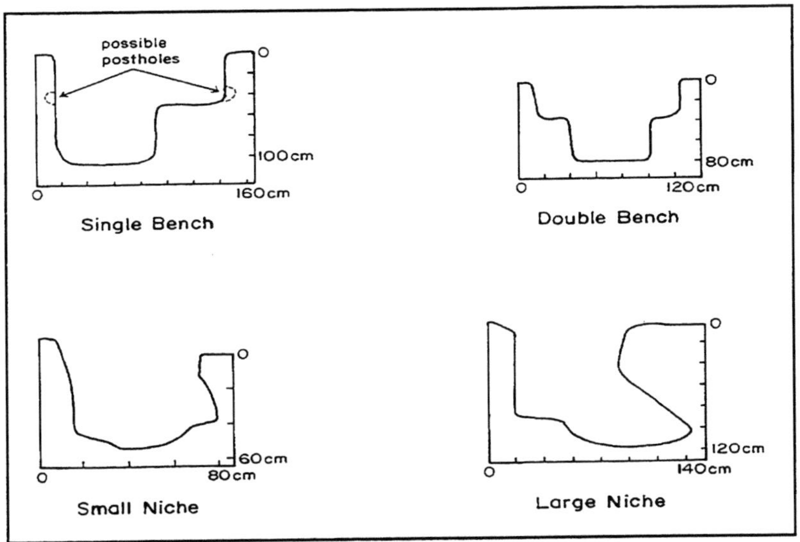

Hohokam grave types at Pueblo Grande. Archaeologists do not know why the Hohokam built different grave forms. Each type may have had some social or symbolic meaning, or represented membership in a particular social group.

crematorium pit, the burned remains of a deceased Hohokam individual were gathered and placed in a ceramic jar or bowl, then buried in a small pit or trench. Occasionally, an inverted bowl or large pottery sherd was placed over the top of the vessel and acted as a cover. In a few cases, more than one vessel containing cremated bones are found together in clusters, inside the same pit.

Tabular schist or slate palettes were sometimes placed with cremation vessels in Hohokam sites. These palettes are rectangular in shape, thin, very flat, and exhibit decorations along the outside border. Many palettes have animal effigies engraved into them. The inner area of the palette generally consists of a rectangular shaped depression.

Analysis of palettes has led archaeologists to argue that during the cremation ceremony, a substance may have been burned in the recessed section of the palette. Lead deposits, probably from the mineral galena, are sometimes found on the palettes and may have given off a bright blue-red fire when burned. It has been suggested that the fire from the palette was symbolic of the doorway leading into the underworld. Palettes appear to have also served other functions. Some show evidence of striations and contain the residues of pigments. These palettes may have been used in the processing of paints.

During the Classic period, Hohokam burial practices changed from predominantly cremating their dead to inhumation, which is the method of burying a body in a grave after death. During this time the practice of cremation continued, but to a lesser extent when compared with earlier periods. This change in burial practices may suggest a change in cultural ideology due to internal changes or external influences. It also may indicate the presence of different ethnic groups living among the Hohokam, or an unwillingness to expend valuable fuel on death rites.

The Hohokam buried their dead in shallow pits dug into the hard desert soil or in softer trash mounds. The exact nature in which a body was interred may have had ceremonial or religious significance for the Hohokam. Upon burial, bodies predominantly were placed in one of two positions, the extended supine position (lying on the back), or on the side in the flexed or fetal position with knees pulled up toward the chest. Heads were commonly faced to the east, toward the rising sun.

At Pueblo Grande, over 700 inhumation burials have been excavated. These inhumations consisted predominantly of the extended supine type, but included a variety of other body positions such as bundle, flexed, sitting, and extended prone (lying face down). Ten inhumation burials have been excavated from the Pueblo Grande platform mound and compound but, interestingly, those burials did not receive more grave goods than burials located elsewhere within the village. Most burials at Pueblo Grande are oriented on an east-west axis with the person's head facing toward the northeast to southeast horizon, possibly indicating the season of interment, but other directions are represented as well.

Most inhumation graves at Pueblo Grande are sub-rectangular shaped, but there were also a small number of oval and circular grave pits. The sub-rectangular pits were of three types: (1) straight-sided, (2) with one or two benches, and (3) niche graves with the body sometimes placed in the niche. The bench burial types often were covered with wood (e.g., saguaro ribs) or brush.

The Hohokam placed many kinds of funerary objects with their dead, for both sexes and for all ages. At Pueblo Grande, evidence has been found of clothing or burial wraps on or around the body. Ceramic pots often surround the body and may have contained offerings of food and water or other perishable material. Shell rings, bracelets, and shell and turquoise necklaces adorned the body. In some burials, the person's head was painted with red, blue-green, or yellow colors. Other funerary items include bone hair pins and awls, projectile points, quartz crystals, spindle whorls, and various minerals. Some individuals appear to have been buried with the tools of their trade.

A burial northeast of the Pueblo Grande platform mound contained unusual funerary objects, suggesting that the individual, a male, was a priest or shaman. Among the various objects placed in the benched grave were three quartz crystals, several shell ornaments, five ceramic vessels, bone hairpins and awls, an obsidian projectile point, four golden eagle wings, and three wings and a lower leg of a raven. Eagles and ravens are considered very sacred birds by many modern Southwestern American Indians. One of the shell ornaments in this burial was shaped into a lizard design, perhaps representing membership in a particular social or religious group.

Studies of inhumation skeletal remains have revealed much information about the Hohokam in the Classic period which was not available for earlier periods. For ex-

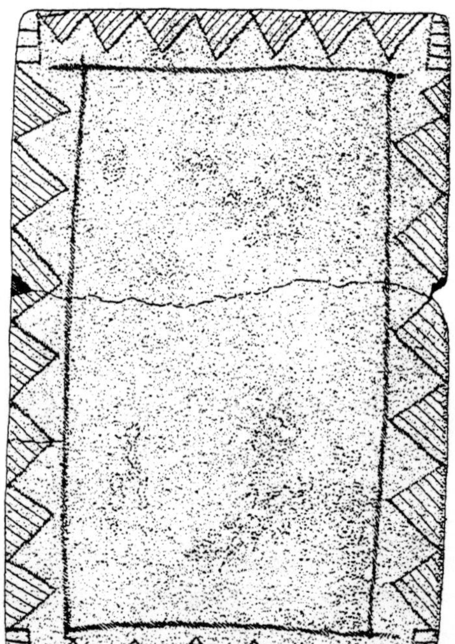

Incised stone palette from Pueblo Grande. Palettes were commonly used in the Colonial period as part of Hohokam cremation ceremonies. A substance burned on the palette may have given off smoke or colored fire representing a door into the underworld.

ample, the average Hohokam male was about 5 feet 5 inches (1.7 m) tall, and the average female was about 5 feet 1 inch (1.5 m). It has also been determined that individuals had a life expectancy of around 40 years or less and infant mortality rates were high. Hohokam skeletal remains indicate that they suffered from some of the same health problems as modern people: dental cavities, arthritis, urinary bladder stones, iron deficiency, and general malnutrition.

Archaeological excavations have revealed that the Hohokam established formal cemeteries within their villages. Occasionally, burials were interred under house floors. The cemeteries are usually located near pithouse clusters or residential compounds, suggesting that the cemeteries were for interring individuals who had once lived in those residential areas. More than a dozen cemeteries have been found at Pueblo Grande.

Rock Art

The Hohokam created a variety of rock art known today as *petroglyphs* and *pictographs*. Petroglyphs are images pecked into the rock by hammerstones and are the most common type of rock art produced by the Hohokam. Pictographs, on the other hand, are painted images on rock surfaces. Because pictographs are painted, they tend not to last as long as

Hohokam petroglyph panel in the South Mountains south of Pueblo Grande. Waterbirds are common motifs in Hohokam petroglyphs in the South Mountains as well as on their red-on-buff pottery.

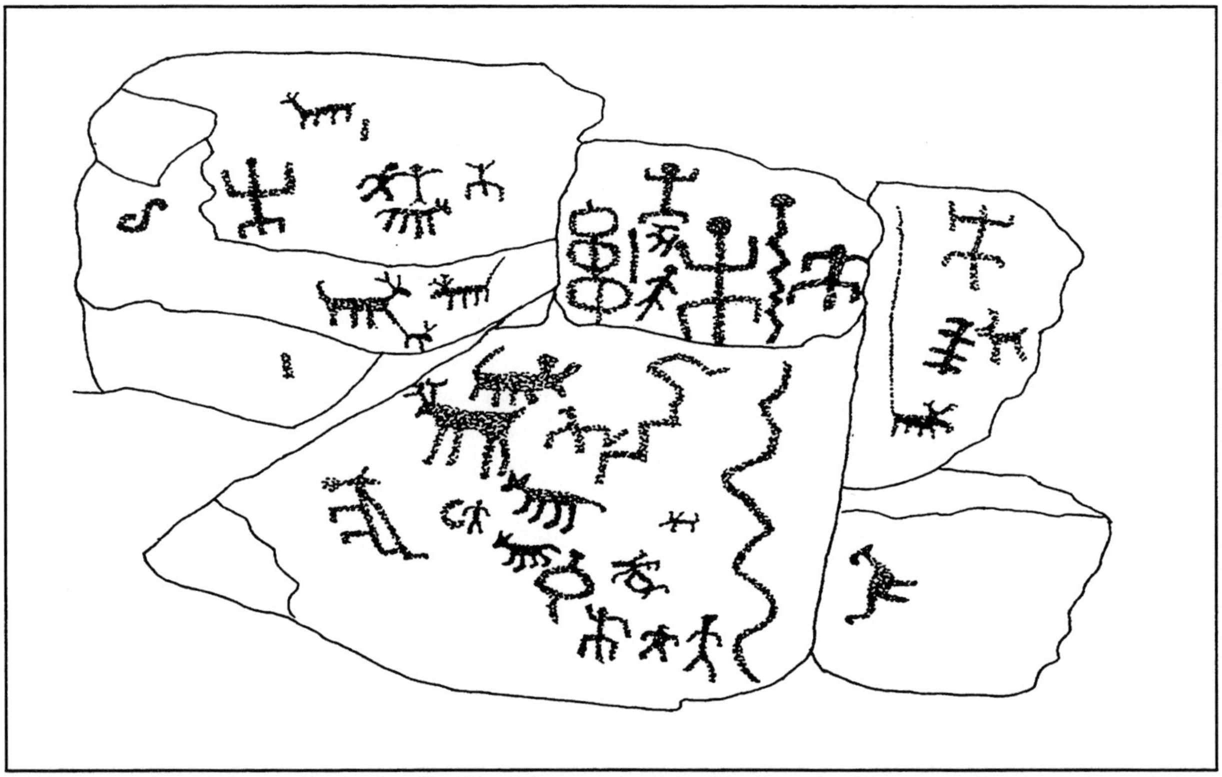

Hohokam petroglyph panel in the South Mountains south of Pueblo Grande. Human, animal, snake and other designs are present. This rock art panel is located at a spring.

petroglyphs and may wear away from exposure to sun, wind, rain, and other elements. This may account for the low number of Hohokam pictographs that have been found.

All of the mountain ranges surrounding the Salt River Valley contain Hohokam petroglyphs, especially the South Mountains, located 5 miles (6.5 km) south of Pueblo Grande. Hohokam petroglyphs also occur in the McDowell Mountains, the Superstition Mountains, the White Tank Mountains, and in the Phoenix Mountains.

The Hohokam used a variety of images and designs in their rock art, both geometric and representational, many of which also appear on their red-on-buff pottery. Geometric designs include circles, spirals, crosses, meandering lines and other elements. Representational images include animals, birds, snakes and humans. Many of the petroglyphs of humans show them performing certain activities such as dancing, flute playing or hunting. Some humans hold objects in their hands and are wearing headdresses. Geometric and representational images are often found mingled together and may represent scenes of some kind.

For many years archaeologists have debated and proposed theories on the meaning and use of rock art. Hohokam rock art may have been *mnemonic* devices (meaning "assisting memories") used to record events, similar to Pima calendar sticks. Some rock art may be clan or totemic symbols, trail markers, boundary markers or territorial signals. Many archaeologists today think that rock art had some kind of ceremonial or religious significance. Petroglyphs and pictographs may have marked sacred locations, and possibly

Hohokam rock art panel from the South Mountains. A lizard, quadraped, snakes and other designs surround two groups of four human figures who appear to be dancing. Many Pima ceremonies involve dancing and singing, often with four repetitions or cycles.

were created as part of priestly or shamanistic rituals such as curing ceremonies or vision quests. Some of the humans, birds, and animals depicted in Hohokam rock art have prominent roles in Pima mythology.

Although no one can be certain what Hohokam rock art meant to its creators, archaeologists have concluded that it was not simply a form of graffiti. Rock art took a considerable amount of time and energy to produce, and some panels of rock art are in practically inaccessible areas high on cliff walls or in caves. This would suggest that the rock art left by the Hohokam was something more than just a means of keeping hands busy, but had a special significance for them.

There is much yet to learn about Hoho-

Hohokam geometric petroglyph design. The "pipette" symbol, shown here, may be a prehistoric rock art design unique to the Hohokam since it is found primarily in Hohokam territory. It sometimes has a face-like appearance that resembles the Mesoamerican rain god, Tlaloc.

kam rock art and further study is likely to shed new insights into this fascinating subject. That is why rock art sites must be protected against vandalism and the encroachment of urban development.

Conclusions

The construction of platform mounds, ballcourts and big houses, as well as the creation of imaginative pottery designs and rock art images, suggest that the Hohokam had an elaborate ideology and worldview. Mythological stories may have been depicted in some of the rock art and pottery designs.

The cycles of nature probably played an important role in the Hohokam belief systems, as they did in Mesoamerica. Archaeologists have determined that the Hohokam had a knowledge of astronomy. Not only did they make observations of the movements of the sun, the moon and possibly other celestial phenomena, but they built architectural features that were used to predict these events. Knowledge of celestial events and phenomena were probably important for regulating ceremonial cycles throughout the year.

Classic period red-on-buff ceramic vessel from the Pueblo Grande collection. The cross design on this jar may represent stars. Archaeologists have determined that the Hohokam were keen astronomical observers based on evidence from architecture and petroglyph sites they left behind.

The Disappearance of the Hohokam?

Footprint petroglyphs from Tempe Butte.

For more than a century, archaeologists have pondered the disappearance of the Hohokam from the Sonoran Desert. Sometime around A.D. 1450, only decades before Columbus arrived in the New World, the Hohokam abandoned their major villages in the Salt and Gila river valleys, including Pueblo Grande. Why, after centuries of successfully cultivating the arid desert, did their society collapse? There have been many competing theories including soil salinization, disease, warfare, floods, droughts, and climatic changes. None or all could be correct; all are still under investigation.

Soil Salinization

Salinization of the soil has been a common explanation for the demise of the Hohokam culture. It has been proposed that the continuous irrigation of their fields over hundred of years caused excessive salt deposits (salinization) to accumulate in the soil, forcing the Hohokam to abandon them. Historically, however, it is known that fields in the Salt River Valley became salinated after a mere 50 years of irrigation. Historic farmers quickly learned to flush their fields with large amounts of water in order to remove salts. They also had to let their fields lie fallow for periods of time to allow the soil to rejuvenate itself.

The Hohokam would have had to deal with this same soil problem soon after settling in the Salt River Valley. The fact that the Hohokam practiced irrigation agriculture successfully for more than 1,000 years indicates that they had addressed this problem. Nonetheless, it is possible that during the Classic period, the Hohokam failed to properly manage their fields as well as they had during previous periods, causing the soil to become too saturated with salt.

Climate: Droughts and Floods

Some archaeologists have proposed that changing climatic conditions forced the Hohokam to move elsewhere. Analysis of tree-ring and stream flow data have shown that the Hohokam experienced massive floods in A.D. 1358 and again in the early 1380s, which in all probability damaged their canal irrigation systems. These floods appear to have been interspaced with periods of drought. But the Hohokam had

Hohokam preparing for war. Recently archaeologists have been taking a new look at warfare throughout the prehistoric Southwest. Although there is little evidence to support Hohokam warfare, Pima legends describe how they rose up against the Hohokam and took over the Hohokam platform mounds by force.

built and repaired canals for hundreds of years. They had experienced flooding in the past, such as a major flood in A.D. 899, and had recuperated from its devastating consequences. Why were they not able to recover from the floods of the 14th century?

Warfare

Other theories have suggested that warfare and raids by other groups, including the Apache, led to the disappearance of the Hohokam from the desert. Other than the existence of a series of possible hilltop fortifications, little evidence has arisen out of the archaeological record to support this idea. Furthermore, the timing of the initial appearance of the Apache and Navajo in the Southwest is still unclear but appears to be well after the Hohokam cultural collapse.

Disease

Another factor that has been proposed for the collapse of the Hohokam involves diseases. During the mid-16th century, the Spanish brought diseases from Europe that killed hundreds of thousands of American Indians in Mesoamerica and the eastern United States. But the Hohokam cultural collapse appears to have occurred before the Spanish arrived in the New World. On the other hand, if local plagues or water borne diseases were introduced into the Hohokam's enormous canal system, the diseases could have spread quickly among the close-knit villages. However, no clear evidence of disease has been found, such as mass graves or large cremation areas, indicating that the Hohokam experienced epidemics.

Migration and Overpopulation

Some archaeologists have suggested that as a result of declining rainfall, neighboring cultures from the north and east abandoned their homes and settled among the Hohokam during the 1300s. This increased population would have dangerously overextended the available resources that the Hohokam had kept in balance for generations. The population of Pueblo Grande probably reached its peak in the Classic period, with as many as 1,000 people living in the village. Studies of burials excavated from the site during this time period indicate that the population was suffering from malnutrition despite their utilization of all available resources, including the intensive harvesting of fish from the Salt River and local irrigation canals. This evidence suggests that the Hohokam at Pueblo Grande may have exceeded the carrying capacity of the desert which, combined with other factors, eventually forced the Hohokam to abandon the village.

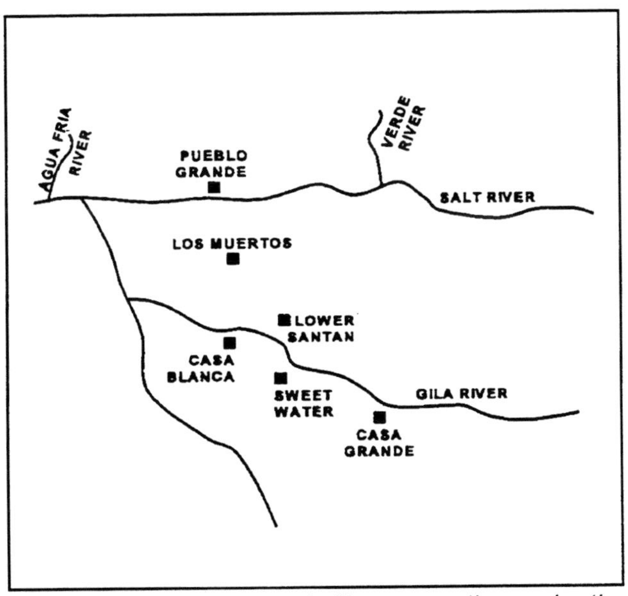

Hohokam platform mound villages overthrown by the Pima as identified in their oral traditions. Pueblo Grande on the Salt River was said to have been one of the last Hohokam villages to succumb to the Pima.

The Hohokam Today

A number of archaeologists argue that although the Hohokam society did indeed collapse, the Hohokam never completely abandoned the Sonoran Desert. Instead, they believe that the American Indian groups living there at the time of the Spaniards' arrival in central Arizona, around A.D. 1694, are the descendants of the Hohokam. Many ancient legends of the Pima (Akimel O'odham) frequently mention the Hohokam. In one version of the Pima oral traditions, a Pima revolt, led by their culture hero Elder Brother, destroyed all the Hohokam platform mound villages. Because Pueblo Grande was so powerful, the Pima had to enlist the aid of shamans and one of them used his power of thunder to destroy the village.

Today, it is generally believed that the Pima and the Tohono O'odham (Papago) of southern Arizona are descendants of the Hohokam. However, oral traditions of the Hopi indians of northern Arizona state that a number of Hopi clans once lived in southern Arizona before migrating to their present location. Furthermore, some archaeologists have noted similarities between the Hohokam culture and Yuman groups in southwestern Arizona.

Conclusions

If the Hohokam never left, why then did they abandon most of their villages in the Salt and Gila river valleys as well as the Tucson Basin, including Pueblo Grande, around A.D. 1450? Perhaps the most plausible theory is that Hohokam society collapsed through internal conflicts triggered by environmental pressures on a population that had met or exceeded the carrying capacity of the land. Archaeologists estimate that there may have been from 24,000 to more than 50,000 Hohokam people living in the Sonoran desert before the Hohokam culture collapsed. The floods of the late-14th century may have severely damaged the Hohokam canal systems which supported such large numbers of people in an arid environment. This, in turn, weakened the control and authority of the secular or theocratic elite. Pima oral traditions tell of how their forefathers overthrew the rulers of the platform mound villages, mentioning Pueblo Grande specifically, because they had grown arrogant. This is surely a sign of social unrest.

With new archaeological evidence coming to light almost daily, it is now known that the Hohokam did not simply vanish or abandon their villages overnight, but instead their cultural collapse appears to have been an extended process lasting several generations. But whatever the cause and however long it took, more than 1,000 years of occupation at Pueblo Grande came to an end in the mid to late 15th century.

Artist's reconstruction of the Pueblo Grande platform mound as viewed from a nearby compound, looking southwest.

Reflections...

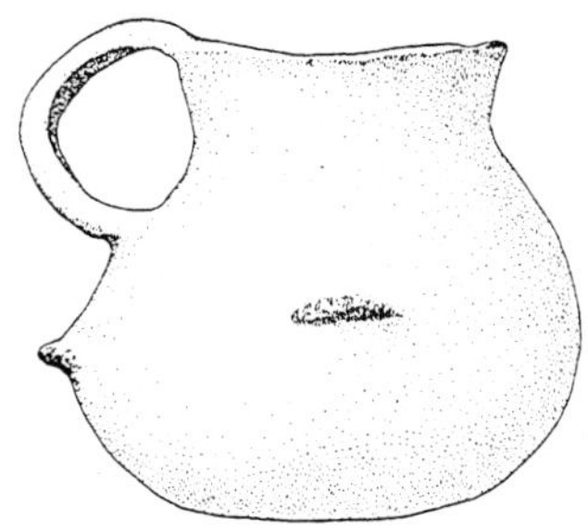

Plain ware vessel with bird wings and tail from Pueblo Grande.

Archaeologists may never know the exact reasons why the Hohokam culture collapsed after more than a millennium of successful adaptation to the Sonoran Desert, but the Hohokam left an indelible legacy behind for future generations to ponder. The village of Pueblo Grande is one of the last preserved Hohokam villages, and it offers an astonishing look at prehistoric Hohokam society.

Pueblo Grande first appeared in the archaeological record ca. A.D. 500, a few hundred years later than some other villages in the Salt River Valley. Over the next ten centuries Pueblo Grande grew from a small hamlet of a few houses to a village of over 1,000 residents. Although Pueblo Grande was only the second largest village along Canal System Two (the site of Las Colinas was larger), its location at the headwaters of the canal system suggests that it would have had political influence among the other villages of the same canal system.

The large platform mound, multiple ballcourts and the existence of the big house suggest that Pueblo Grande held importance as a Hohokam ceremonial center for the entire Salt River Valley. The massive platform mound with its restricted public access, multiple rooms, and ceremonial features suggest the mound and its compound had a variety of functions that consisted of both sacred and secular activities. A small group of priests may have lived on top of the mound and controlled Hohokam religious practices and other ceremonial activities so important to Hohokam society. The priests and other leaders of the Hohokam, however, were unable to prevent, and may have contributed to, their culture's eventual collapse and abandonment of the village of Pueblo Grande around A.D. 1450.

Over a century of archaeological investigations have been conducted at Pueblo Grande which have revealed a wealth of information on the Hohokam (see Appendix 1). The archaeological ruins of Pueblo Grande are a testament to the cultural achievements of these people. Through archaeology we know that the Hohokam were an artistic, inventive, complex, and resourceful people who were tied closely to the land on which they lived. They built large villages containing ballcourts, massive platform mounds, and residential compounds. To survive in the arid Sonoran Desert, the Hohokam built hundreds of miles of irrigation canals which fed thousands of acres of rich agricultural fields. They were connected to their neighbors through thousands of miles of trade routes and were influenced by the civilizations of Mexico and the Southwest. How they viewed the world around them can be seen through their architecture, crafts, rock art, and in the way they cared for their dead.

There is still much to learn about the Hohokam culture, and Pueblo Grande is in a unique position to answer some of the remaining questions. Pueblo Grande Museum and Archaeological Park was created in part for this purpose and is dedicated to the study, proper development, and preservation of this magnificent archaeological site. (For a history of the Museum and Park see Appendix 2.) Through continued preservation and careful study of Pueblo Grande, many aspects of over 1,000 years of Hohokam cultural history will be available for future generations to appreciate.

Entrance to the Pueblo Grande Museum. Today's museum interprets the Hohokam village ruins on-site and houses exhibits on the Hohokam people and the science of archaeology. It is also the repository for archaeological investigations conducted at the Pueblo Grande site and other sites in the City of Phoenix.

Appendix 1

Archaeological Investigations at Pueblo Grande	
Date	**Person/Organization**
1883	Adolph Bandelier/Archaeological Institute of America
1887	Frank Hamilton Cushing/Hemenway Southwestern Archaeological Expedition
1901	Dr. Joshua Miller/Arizona Antiquarian Association
1925	Erich F. Schmidt/American Museum of Natural History
1929-1934	Odd S. Halseth/City of Phoenix
1934-1935	James W. Simmons/Public Works Administration
1935-1936	Odd S. Halseth/Civilian Conservation Corps
1936-1940	Julian D. Hayden/Civilian Conservation Corps
1938-1940	Albert Schroeder, Audie Kelley and Paul Ezell/Works Progress Administration
1948	Donald P. Jewell/City of Phoenix
1953-1956	Donald H. Hiser/City of Phoenix
1958-1973	Donald H. Hiser/City of Phoenix
1959	Richard Woodbury/University of Arizona Arid Land Program
1969	Walter T. Duering/City of Phoenix
1975	Jon S. Wood, David E. Ward, and Robert E Gasser/City of Phoenix
1977-1980	Donald H. Hiser and Chad Phinney/City of Phoenix
1985	David A. Gregory/Arizona State Museum and City of Phoenix
1985	John S. Cable and David E. Doyel/City of Phoenix
1989-1990	Robert I. Birnie/City of Phoenix
1989-1990	Christian E. Downum and James P. Holmlund/City of Phoenix
1989-1990	John S. Cable and Michael S. Foster/Arizona Department of Transportation
1990	Jerry B. Howard and Todd W. Bostwick/City of Phoenix
1990	Robert I. Birnie and Mary Ellen Walsh-Anduze/ City of Phoenix
1995	K. J. Schroeder/City of Phoenix
1999-2000	Soil Systems, Inc.

Aerial photograph of original Pueblo Grande Museum and platform mound, ca. 1940, view looking toward northeast.

Appendix 2

The Advent and Growth of the Pueblo Grande Museum and Archaeological Park

Date	Event
1924	Thomas Armstrong donates platform mound and approximately five acres (2 ha) of surrounding area to the City of Phoenix.
1924	Omar Turney, City of Phoenix Engineer, convinces city to purchase the Park of Four Waters, approximately 10 acres (4 ha), south of the platform mound.
1929	Odd S. Halseth is hired as the first City of Phoenix Archaeologist and director of Pueblo Grande Museum. First city archaeologist in the nation.
1933-1934	First Museum building constructed of adobe bricks west of platform mound.
1936	Approximately three acres (1 ha) of land west of platform mound purchased.
1951	Approximately 11 acres (4 ha) of land north of platform mound purchased from Tovrea Stockyards.
1960	Donald Hiser replaces Odd Halseth as City of Phoenix Archaeologist and director of Pueblo Grande Museum.
1963	Park of Four Waters is designated a National Historic Landmark.
1964	The platform mound and surrounding area are designated as a National Historic Landmark.
1968-1986	Approximately 72 additional acres (29 ha) of land purchased, bringing total acreage of Pueblo Grande Museum and Archaeological Park to 102.27 acres (41 ha).
1974	New Museum building constructed, replacing old adobe one.
1974	Pueblo Grande is combined into a single National Historic Landmark.
1977	Pueblo Grande Museum Auxiliary founded.
1977	Pueblo Grande Museum holds its first Indian Market.
1984	Dr. David E. Doyel becomes City of Phoenix Archaeologist and director of Pueblo Grande Museum.
1988-1990	National Park Service evaluates eroded condition of Pueblo Grande ruins and conducts stabilization of the platform mound and ballcourt.
1989	Pueblo Grande Museum Archival Project begun.
1990	Roger W. Lidman becomes director of Pueblo Grande Museum and Todd W. Bostwick becomes City of Phoenix Archaeologist.
1995	Two additional buildings added to Museum complex, permanent collections storage and a community room, doubling the size of the Museum's facilities.

The first Pueblo Grande Museum building, looking northwest with the Grand Canal in foreground, 1939. The first museum building was built of adobe bricks made on site. It was completed in 1934 and served for nearly 40 years as an educational institution and repository of archaeological material excavated from the Pueblo Grande site as well as several other sites in the Salt River Valley. This building was replaced by the current museum building in 1974.

Recommended Reading

Abbott, David R., ed.
1994 *The Pueblo Grande Project, Vol. 3: Ceramics and the Production and Exchange of Pottery in the Central Phoenix Basin.* Soil Systems Publications in Archaeology No. 20, Phoenix.

Bostwick, Todd W.
1998 *Hohokam Rock Art: Ancient Images Left in Stone.* Pueblo Grande Museum Profiles No. 18, Phoenix.

Bostwick, Todd W. and Christian E. Downum, eds.
1994 *Archaeology of the Pueblo Grande Platform Mound and Surrounding Features, Vol. 2: Features in the Central Precinct of the Pueblo Grande Community.* Pueblo Grande Museum Anthropological Papers No. 1, Phoenix.

Crown, Patricia L. and James W. Judge, eds.
1991 *Chaco & Hohokam: Prehistoric Regional Systems in the American Southwest.* School of American Research Press, Santa Fe.

Downum, Christian E. ed.
1998 *Archaeology of the Pueblo Grande Platform Mound and Surrounding Features, Vol. 4: The Pueblo Grande Platform Mound Compound.* Pueblo Grande Museum Anthropological Papers No. 1, Phoenix.

Downum, Christian E. and Todd W. Bostwick, eds.
1993 *Archaeology of the Pueblo Grande Platform Mound and Surrounding Features, Vol. 1: Introduction to the Archival Project and History of Archaeological Research.* Pueblo Grande Museum Anthropological Papers No. 1, Phoenix.

Doyel, David E., ed.
1987 *The Hohokam Village: Site Structure and Organization.* American Association for the Advancement of Science, Glenwood Springs, Co.

Foster, Michael S., ed.
1995 *The Pueblo Grande Project, Vol. 4: Material Culture.* Soil Systems Publications in Archaeology No. 20, Phoenix.

Gumerman, George J., ed.
1995 *Exploring the Hohokam: Prehistoric Desert Peoples of the American Southwest*. University of New Mexico Press, Albuquerque.

Hamann, Diane L.
1995 Hohokam Rock Art of Southern Arizona: A Regional Overview. In *Rock Art Studies in the Americas*. Edited by Jack Steinbring, pp. 77-82. Oxbow Monograph No. 45, Oxford.

Haury, Emil W.
1976 *The Hohokam: Desert Farmers and Craftsmen.* University of Arizona Press, Tucson.

Howard, Jerry B. and Gary Huckleberry
1991 *The Operation and Evolution of an Irrigation* System: *The East Papago Canal Study.* Soil Systems Publications in Archaeology No. 18, Phoenix.

Kwiatkowski, Scott, ed.
1994 *The Pueblo Grande Project, Vol. 5: Environment and Subsistence.* Soil Systems Publications in Archaeology No. 20, Phoenix.

Masse, W. Bruce
1976 *The Hohokam Expressway Project: A Study of Irrigation in the Salt River Valley, Arizona.* Contributions to Highway Salvage Archaeology in Arizona No. 43. Arizona State Museum, Tucson.

Mitchell, Douglas R., ed.
1994 *The Pueblo Grande Project, Vol. 2: Feature Descriptions, Chronology, and Site Structure*. Soil Systems Publications in Archaeology No. 20, Phoenix.

Noble, David G., ed.
1991 *The Hohokam: Ancient People of the Desert.* School of American Research Press, Santa Fe.

Scarborough, Vernon L. and David R. Wilcox
1991 *The Mesoamerican Ballgame*. University of Arizona Press, Tucson.

Wilcox, David R., Thomas R. McGuire and Charles Sternberg
1981 *Snaketown Revisited.* Arizona State Museum Archaeological Series No. 155, Tucson.

Wilcox, David R. and Charles Sternberg
1983 *Hohokam Ballcourts and Their Interpretation.* Arizona State Museum Archaeological Series No. 160, Tucson.

Acknowledgments

The authors would like to thank many people who assisted with this book. We greatly appreciate help from Roger Lidman, Pueblo Grande Museum Administrator for his support and encouragement; Pueblo Grande Museum staff Laura Andrew, Tom Brewer, Glena Cain, Frank DeCurtis, Mike Fulghum, Gerri Green, Pam Groff, Thomas Hulen, Wilmer Joshevama, Kraig Kemp, Rob Kent, Esteban Linares, Barbara Moulard, Sandy Oglesby, Cathy Reigle, Steve Ross, Robert Serocki, Isabel Worden-Klym, Holly Young; Pueblo Grande Museum Auxiliary members Beverly Schueneman, Ed Grobe, Mary Grobe, Courtney Hammond, Lyda Harris, and Clifford Straitor; and our wives Christina Andrews and Heidi Bostwick for their comments and suggestions.

Information about the Pueblo Grande Hohokam Expressway Project was obtained from Soil Systems, Inc. — Cory Breternitz, Michael Foster, Douglas Mitchell, and David Abbott — and from Bettina Rosenberg, Arizona Department of Transportation. Christian Downum, Director of the Anthropology Laboratories at Northern Arizona University, provided his interpretation of the architectural construction sequence of the Pueblo Grande platform mound.

Illustrations were done by John Andrews and David Morris (page 14), Todd Bostwick (pages 4, 11, 16, 36 [bottom], 37, 48, 58, 62, 63, and 65), Glena Cain (pages 18, 19 [bottom], 20, 21, 22, 27, 36 [top], 44, 54, 55 [top], and 68), Philissa Calamity (pages 47 and 67), Allison Dean (pages vi, 2, 10, 13, 24, 26, and 59), Christian Downum (page 30), Dan Gronseth (page 7), Michael Hampshire (cover), John Jennet (pages 29, 32, 34, and 57 [bottom]), Jonathan Joha (pages 1, 3, 9, 19 [top], 25, 38, 39, 40, 43, 45, 52, 53, 55 [bottom], 56, 57 [top], 60, 66, 69 and 79), and David Morris (page 15).

Photographs were taken by Laura Andrew and John Andrews (pages 41 [bottom], 42, 44, 46, 47, 48, 49, 50 [top], and 51), Laura Andrew (page i), John Andrews (page 8), Todd Bostwick (pages 12, 34, 50 [bottom], 61, and 64), Glena Cain (page 41[top]), Roger Lidman (page 40), Sandy Oglesby (page 17), and Bob Rink (pages 28, 29, 33, and 70).

Funding for the illustrations and photographs came in part from the Pueblo Grande Museum Auxiliary. Formatting of this book was done by John Andrews. The City of Phoenix Parks, Recreation and Library Department provided financial support for the publication of the book.

Glossary

Adobe: Sun-dried puddled mud or bricks made of dirt and clay. (The Hohokam did not make adobe bricks.)

Anthropomorphic: Having the form or shape of a human (e.g., figurines and rock art).

Archaic: Archaeological term for ancient hunters and gatherers.

Artifact: Any object fashioned or used by humans.

Band: A small group of humans, usually 25 people or less.

Beringia: The prehistoric land bridge between Siberia and Alaska.

Caliche: A rock-like calcium carbonate material found in arid regions, deposited in layers and created by the filtering of water through desert soil.

Ceramics: Objects of fired clay such as pottery, pipes, and figurines.

Coursed-Adobe: The placing of adobe mud into layers for construction purposes.

Ethnography: Using the knowledge gained by the study of living people to infer the use of an object used by a prehistoric people.

Feature: Similar to an artifact, a feature was made or produced by humans, but unlike an artifact, it cannot be removed without destroying it. Features are things such as houses, fire pits, storage pits, ballcourts, or platform mounds.

Horno: Spanish for "oven;" often used to describe Hohokam pits used to bake foods.

Hunter-Gatherers: People who depend on the hunting of wild animals and the gathering of wild plants for food to survive.

Lithic: The word lithic comes from the Greek word "lithos," meaning stone. It is one of the materials that past cultures used to make tools.

Mano: A hand-held stone used with a metate to grind dried corn and seeds.

Megafauna: Big game animals of the Late Pleistocene or last Ice Age hunted by Paleo-Indians. These were animals such as mammoths, mastodons, giant sloths, and saber tooth cats.

Metate: A base stone with a flat or depressed surface on which grains and seeds are ground using a mano.

Obsidian: A naturally occurring volcanic glass used to make tools by prehistoric people.

Olla: A large ceramic pot, usually used to hold water.

Paleo-Indians: Meaning ancient. Considered by many archaeologists to be the first American Indians to have migrated into the Americas at the end of the Pleistocene. Often called "Big-Game Hunters."

Park of Four Waters: The area along the Salt River, south of the Pueblo Grande platform mound, containing the preserved remains of Hohokam irrigation canals.

Petroglyph: A design pecked into rock.

Pictograph: A design painted on rock.

Pithouse: A common type of prehistoric house built by the Hohokam and other Southwestern cultures. This type of house is distinct by its placement in a shallow pit (1 to 1 ½ feet deep).

Pleistocene Epoch: The Ice Age(s). It is the time period from approximately 1.6 million years ago to 10,000 years ago. It is characterized in North America by periods of glacial advance and retreat, respectively called glacials and interglacials.

Polychrome: A ceramic pottery vessel painted with three or more colors.

Projectile Point: A term used by archaeologists to describe arrowheads, spearheads and dart points.

Pueblo Grande: Spanish for *Large Town*.

Ramada: Spanish for *shade*. A ramada is a covered outdoor activity area with open sides.

Rock Art: The pecking, incising, or painting of designs onto rock surfaces.

Salinization: The accumulation of excessive salts in soils due to water infiltration from rainfall or during irrigation.

Sedentary: Living in one place year round.

Sherd: A fragment or piece of a ceramic vessel (pottery).

Slip: A thin outer covering of clay on a ceramic pot, generally used to create a background for painted designs.

Site: An archaeological term for a place where humans lived, performed some type of activity or left their artifacts behind.

Temper: Coarse material such as sand, crushed rock, crushed sherds, or shell added to fine clay to make it bond during firing.

Tinklers: Small bell-like objects made from marine shell.

Tumpline: A strap worn around the forehead used to support the weight of heavy objects carried on the back.

Zoomorphic: Having the form or shape of an animal (e.g., figurines and rock art).

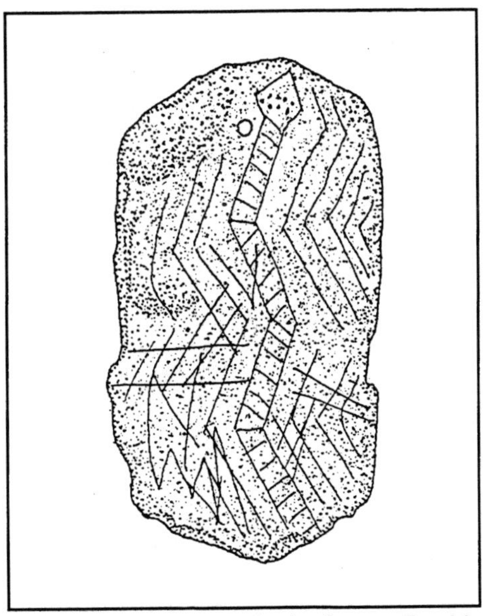

Schist stone pendant from Pueblo Grande. A snake and lightning design have been incised into this pendant.

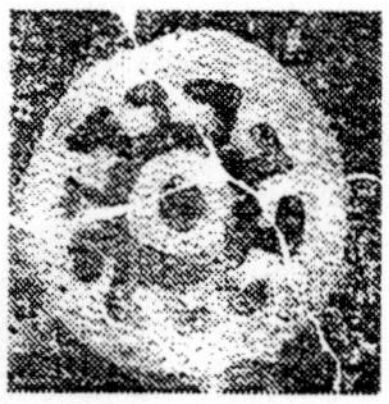

Pueblo Grande Museum

and Archaeological Park

City of Phoenix Parks, Recreation and Library Department
4619 East Washington Street
Phoenix, Arizona 85034
Ph: (602) 495-0901 * Fax: (602) 495-5645
www.pueblogrande.com

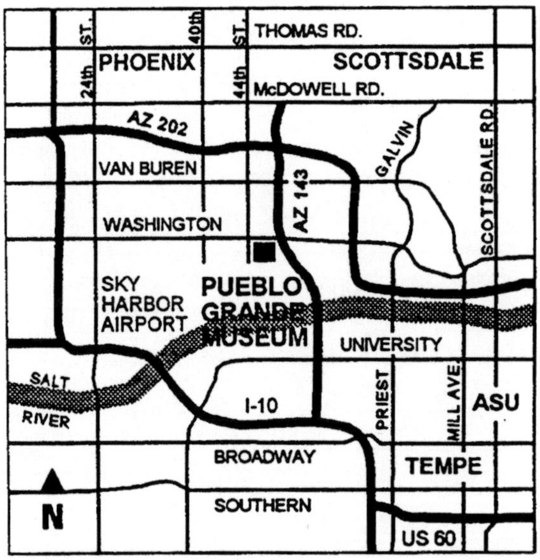